THE CELL CYCLE
an introduction

THE CELL CYCLE
an introduction

ANDREW MURRAY
University of California,
San Francisco

TIM HUNT
ICRF Clare Hall Laboratories
Hertfordshire, UK

W. H. FREEMAN AND COMPANY
New York

Library of Congress Cataloging-in-Publication Data

Murray, Andrew Wood.
    The cell cycle : an introduction / Andrew Murray, Tim Hunt.
      p.   cm.
    Includes index.
    ISBN 0-7167-7044-X (hard). — ISBN 0-7167-7046-6 (soft)
    1. Cell cycle.   I. Hunt, Tim, 1943–   . II. Title.
QH605.M95   1993
574.87′623 — dc20   93-10477
                                                           CIP

Printed in the United States of America

1 2 3 4 5 6 7 8 9 0   HP   9 9 8 7 6 5 4 3

*Dedicated to Anna Murray
and to the memory of
Kit and Richard Hunt
and Robert Murray*

CONTENTS

PREFACE

How do cells grow and divide? This question has fascinated biologists for over a century and is directly relevant to cancer and other important medical problems. In talking with colleagues and students we realized that although most of them knew that our understanding of the cycle of growth and division had advanced dramatically, many wished to know how knowledge derived from many different organisms had been assembled to create a unified picture of the cell cycle. This book attempts to describe the origin of the different experimental approaches to the cell cycle and traces their contribution to our current picture of cell growth and division.

We attempt to reach a wide audience, ranging from undergraduates taking advanced courses in the biological sciences to researchers working on the cell cycle and related problems. To achieve this goal, we concentrate on a small number of organisms and critical experiments that have advanced our understanding of the cell cycle, rather than trying to produce an encyclopedic review. In addition, we restrict our citations from the scientific literature to the experiments that are depicted in figures and a few important original and review articles that relate to the central themes of each chapter. Finally, we do not (with one exception) cite the names of living scientists in the text. We hope that all those whose work is not discussed or cited and whose names are not mentioned will understand that these omissions were made in the quest for brevity and coherence.

A great many people contributed to the genesis and maturation of this book. We are indebted to Daniel Friend, W. N. Hittleman, Marc Kirschner, Ira Herskowitz, and Eric Schabtach for photographs, and we owe special thanks to N. Hajibagheri of the ICRF electron microscopy unit and Ted Salmon, Bob Skibbens, and John Robinson, who made photographs especially for the book.

We are very grateful to those colleagues who made valuable suggestions on the manuscript as it evolved. Molly Bourne, Chris Ford, R. A. Hyman, Mike Kilgard, Paul Nurse, Jonathon Pines, Peter Sorger, Julie Theriot, and Julia Turner all read an early draft of the entire book. Mike Bishop, Henry Bourne, Webb Cavenee, Willie Donachie, Piet de Boer, Sandra Gerring,

Doug Kellog, Ira Herskowitz, Rong Li, Jeremy Minshull, Tim Mitchison, David Morgan, and Geoff Wahl read particular chapters. Kim Arndt, Breck Byers, Tony Carr, Martha Cyert, Dean Dawson, Michelle Flatters, Bruce Futcher, Kathy Gould, Lee Hartwell, Phil Hieter, Lee Johnston, Ira Herskowitz, Greg May, Ron Morris, Paul Nurse, Steve Osmani, Shirleen Roeder, Paul Russell, Viesturs Simanis, Bruce Stillman, Malcolm Whiteway, Mark Winey, Masayuki Yamamoto, and Mitsuhiro Yanagida read the Appendix. We hope that the helpful suggestions, corrections, and occasional trenchant criticisms made by these readers have transformed the original sow's ear. If not, the fault is entirely ours.

The book was conceived at the Marine Biological Laboratory in Woods Hole, an institution that has played a pivotal role in both our lives. We heartily thank all those who have made summers there such a lively and rewarding experience, especially John Kilmartin, Tom Pollard, Conly Rieder, Joel Rosenbaum, Joan Ruderman, Ted Salmon, Katherine Swenson, Kip Sluder, and Ron Vale. We thank Gary Odell, who introduced us to the wonders of mathematical modeling and constructed the elegant model in Figure 4-22. We are also deeply indebted to our colleagues who have discussed aspects of the cell cycle with us over the years, including Dean Dawson, Marcel Doree, Eric Karsenti, Tim Mitchison, Jim Maller, David Morgan, Kim Nasmyth, Takeharu Nishimoto, Bruce Nicklas, Terry Orr-Weaver, Joan Ruderman, Ted Salmon, Jack Szostak, and Mitsuhiro Yanagida. We are especially grateful to John Gerhart, Lee Hartwell, Marc Kirschner, and Paul Nurse. Their seminal experimental and intellectual contributions sparked our interest in the cell cycle, and discussions with them have directed our thinking. It has been a great pleasure to grow up in the cell cycle with such good friends and in such distinguished company.

We thank all those at W. H. Freeman who bridged the yawning chasm from idea to book. Kirk Jensen persuaded us to embark on the project and Ingrid Krohn provided support and advice at crucial moments. We thank Stephanie Hiebert for cutting through thickets of tangled prose, Megan Higgins for advice about artwork, and Alison Lew for the design of the book. Above all, we thank Penny Hull, our project editor. Without her excellent advice, organization, endless patience, and good humor, this book would have never seen the light of day.

Finally, we beg the indulgence of those across whose lives the shadow of this project has fallen: members of our labs, our friends, our families. We apologize to them for our dereliction of duty and endless procrastination and thank them for their patience and support. Words cannot express our gratitude to Joanne Hackett and Roger Murray for food, drink, and friendship, and to Helen Epstein and Barbara Kramer for love and understanding.

ANDREW MURRAY
TIM HUNT
April 1993

NOTE ON GENETIC NOMENCLATURE

There is no universal convention for naming genes, mutations, and gene products in different organisms. A few standards do exist, however. Names of genes and mutants are always italicized and usually consist of a three-letter abbreviation that is followed by a number or an additional letter. To reduce confusion in this text, we have appended to each gene name one of four superscripts: $+$, $-$, ts, or D. Thus, yfg^+ refers to the wild-type form of "your favorite gene," yfg^- to a recessive mutant, yfg^{ts} to a recessive, temperature-sensitive mutant, and yfg^D to a dominant mutant.

For budding yeast, each three-letter gene name is followed by a number (there is one exception, the HO gene) Names of wild-type genes and dominant mutations are uppercase (e.g., $CDC28^+$ and $CLN3^D$). Names of recessive mutations are lowercase: For example, $cdc4^{ts}$ and $cdc28^{ts}$ are temperature-sensitive mutations of two different cell division cycle (CDC) genes. To specify a particular mutation in a gene, the gene number is followed by an allele number. Thus, $cdc28-4^{ts}$ and $cdc28-13^{ts}$ are two different temperature-sensitive mutations of the $CDC28$ gene. Names of gene deletions also are lowercase and are preceded by "Δ" (e.g., Δ$rad9$).

For fission yeast, all gene designations, whether for wild-type or mutant forms, are lowercase. Thus, $cdc10^+$ is the wild type version, $cdc10^-$ a recessive mutation, $cdc10^{ts}$ a temperature-sensitive mutation, and $cdc10^D$ a dominant mutation of the $cdc10$ gene. The conventions for allele numbers and gene deletions are the same as those for budding yeast.

For bacteria, gene designations are followed by an uppercase letter instead of a number. Thus $dnaA^{ts}$ and $dnaB^{ts}$ (DNA synthesis) are temperature-sensitive mutants in two different genes involved in DNA synthesis, and $dnaA^+$ is the wild-type version, $dnaA^-$ is a recessive mutation, and $dnaA^D$ is a dominant mutation in the $dnaA$ gene. The conventions of genetic nomenclature in the fungus *Aspergillus nidulans* are essentially identical to those for bacteria.

There is no reliable convention for the names of genes in vertebrate cells and in much of the literature the distinction between a gene and its protein product is not made clear by the form of the abbreviation. We have used the convention that gene names are italicized and followed by the same subscripts used for other organisms. In the case of viral genes that have cellular homologs the viral gene name is preceded by *v-* and the cellular gene name by *c-*. Thus *v-src* and *c-src* are the viral and cellular versions of a gene involved in controlling cell proliferation.

Gene products are designated in the same way for all organisms: by using the same abbreviation as for the gene, but with only the first letter capitalized and without italics. Thus, Cdc7 is the protein product of the budding yeast *CDC7* gene, and Cdc10 is the protein product of the fission yeast *cdc10* gene. The corresponding mRNAs would be referred to as the Cdc7 mRNA and the Cdc10 mRNA.

INTRODUCTION

*T*HIS BOOK is about the cell cycle, the ordered set of processes by which one cell grows and divides into two daughter cells. Cell growth and division is a cornerstone of biology. Without understanding how the cell cycle works, we cannot understand how the fusion of two cells, an egg and a sperm, and the subsequent divisions of the fertilized egg produce an adult human composed of about 10^{13} cells. Without knowing the checks and balances that normally ensure orderly cell division, we cannot devise effective strategies to combat the uncontrolled cell divisions of the cancers that will kill one in six of us.

In the last five years there has been a revolution in our understanding of the cell cycle so thorough that this book bears little relationship to its predecessors. This transformation arose in a different way from earlier revolutions in biology. The earlier advances were mostly due to experiments on a single organism, performed by an individual or by members of a closely related school of scientists who shared a common outlook on biology. For example, Gregor Mendel discovered the fundamental principles of genetics by breeding peas in a Hungarian monastery; Thomas Hunt Morgan and his colleagues at Columbia University invented genetic mapping and, by investigating mutations in the fruit fly, *Drosophila melanogaster,* showed that genes were located on chromosomes; and at the Pasteur Institute, François Jacob and Jacques Monod discovered how genes were turned on and off by studying how the bacterium *Escherichia coli* metabolizes lactose.

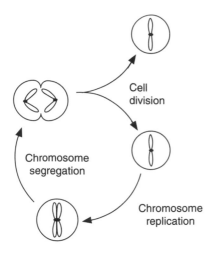

FIGURE I-1   Basic cell cycle. A very schematic view of one cell cycle in an idealized cell with a single chromosome. The cycle comprises three essential events: chromosome replication, chromosome segregation, and cell division. The dark circle on each chromosome is the kinetochore, the protein–DNA complex that attaches the chromosomes to the machinery responsible for segregating the chromosomes to the two daughter cells.

The earlier upheavals in biology each had a single focal point, but the revolution in the cell cycle has depended on the contributions of a diverse array of investigators who studied different organisms and initially had very different philosophies of the cell cycle. For many years scientists actively debated whether the cell cycles of different cells ran in fundamentally different ways. We know now, however, that the cell cycles of all eukaryotes (cells with nuclei) are governed by the same principles. The key proteins that regulate the cell cycle have been so well conserved during evolution that human proteins can successfully regulate the cell cycle of simple yeasts even though our ancestors diverged more than a billion years ago. The cell cycle revolution depended on the advent of genetic engineering. The crucial experiments that unified the different views of the cell cycle would have been impossible without the invention of techniques to clone, sequence, and manipulate genes.

Most cells must complete four tasks during the cell cycle. They must grow, replicate their DNA, segregate their chromosomes into two identical sets, and divide (Figure I-1). We have tried to create an integrated approach to thinking about the cell cycle, rather than proceeding chronologically through its stages. Chapter 1 introduces the basic processes of the cell cycle. Chapters 2 to 4 each focus on a particular cell type that contributed to answering a key question about the cell cycle. These chapters describe the critical experiments that revealed the logic of the cell cycle and attempt to place them in a historical perspective with other important approaches to the same problems. Chapters 5 to 8 elaborate on the initial chapters by providing detailed descriptions of particular processes in the cell cycle. Finally, Chapters 9 to 11 describe the involvement of the cell cycle in cancer and explore the specialized cell cycles involved in sexual reproduction and the growth and division of bacteria. Throughout the book, we try to estimate how far we have come on the road to enlightenment and attempt to identify the most important unanswered questions about the cell cycle.

CELLS AND THE CELL CYCLE

*I*N 1 8 3 8 Theodor Schleiden and Jacob Schwann proposed the cell theory. Their proposal had two main tenets: that every living organism is composed of one or more cells and that new cells can arise only by the division of preexisting cells. Cell division is the only path to immortality. Nondividing cells can live for as long as a hundred years, but they always eventually die. Viruses are the one apparent exception to the cell theory, but since they can replicate only inside cells, their continued existence depends on the continued reproduction of cells.

Since the early part of the twentieth century, scientists have understood that the chromosomes carry genetic information. To reproduce, cells must replicate their chromosomes and then segregate them into two daughter cells. By the late 1950s the demonstration that DNA was the genetic material and the elucidation of its structure enabled biochemists to ask questions about DNA replication. Advances in cell biology during the 1960s identified the molecules that form the **cytoskeleton** that gives cells their shape and organizes the segregation of chromosomes and the division of cells. Thus by 1970, scientists understood the basic composition of cells and could frame questions about the molecular details of individual processes in the cell cycle. This chapter describes the fundamental properties of cells and the cell cycle and introduces a series of experiments that raised fascinating questions about the logical organization of the cell cycle.

BASIC PROPERTIES OF CELLS

Cells are complicated objects of many different types. The most important distinction is between **eukaryotic cells,** whose DNA is enclosed in a **nucleus,** and **prokaryotic cells** (such as bacteria), which lack any distinction between nucleus and cytoplasm (Figure 1-1). This section attempts to identify the universal properties of cells that are relevant to the cell cycle.

A cell is a cooperative of molecules capable of reproducing itself

Cells are discrete entities that grow and divide. To do so, a cell must have a number of components: a set of instructions and the production facilities for making the molecules that will form a new cell, a mechanism for dividing one cell into two, and a defined boundary between itself and its environment across which nutrients are imported and wastes exported.

The instructions are encoded in the sequence of DNA in the **chromosomes.** During the cell cycle, the chromosomes have two functions. As carriers of genetic information, they must be duplicated by specialized replication enzymes and then segregated into two daughter cells as a result of rearrangements in the cytoskeleton that also divide the cell in two. In addition, they must express their genes to direct the production of two classes of proteins: those that regulate the cell cycle and those that form the components of the next generation of cells.

The border between the cell and its environment is defined by the **plasma membrane,** which is composed of a mixture of lipids and proteins. The plasma

FIGURE 1-1 Prokaryotes and eukaryotes. Electron micrographs of a mouse white blood cell and of a bacterium. The human cell contains many discrete membrane-bound compartments, the most prominent of which is the nucleus. By contrast, the bacterium lacks any boundaries within its cytoplasm. Micrographs courtesy of Daniel Friend.

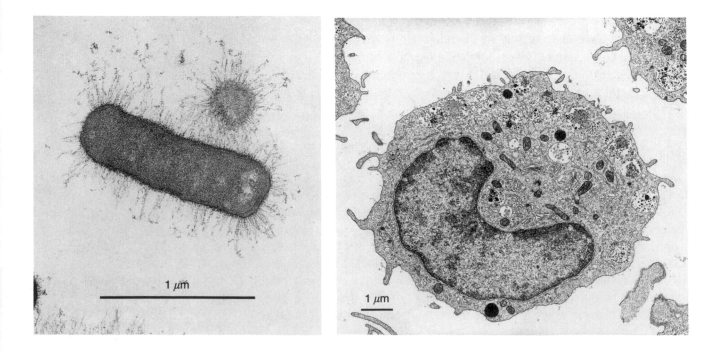

1 µm

1 µm

membrane is impermeable to large molecules and to most of the small molecules inside the cell. Membrane proteins mediate the interactions between the cell and its environment. Transport proteins bring nutrients into cells, export waste products, and maintain the ionic composition of the cytoplasm, and membrane receptors allow cells to respond to chemical signals sent by other cells.

Eukaryotic cells have discrete organelles — including the nucleus, mitochondria, lysosomes, Golgi apparatus, and endoplasmic reticulum — each bounded by one or two membranes. Exchanging materials between organelles involves traffic of membrane vesicles and proteins between organelles, while exchanging proteins between the cell and its environment involves vesicle traffic between organelles and the plasma membrane.

Cells contain thousands of different proteins

How complicated is a cell? A reasonable guess is that a cell needs between 2,000 and 5,000 different enzymes and structural proteins to grow and divide. The amounts of different cell components vary considerably: Some molecules, like ribosomal proteins and RNAs, are present in millions of identical copies per cell; others, like DNA, are present as only one or two.

Cells contain many different types of proteins, each specialized for a particular role in the life of the cell. Important classes include the enzymes that produce the building blocks for the synthesis of DNA, RNA, and proteins and the enzymes that use these building blocks to replicate DNA, transcribe DNA into RNA, and translate mRNA into protein. The form and function of cells depend on the structural proteins that form the cytoskeleton and on the motor proteins that move objects, such as chromosomes, along elements of the cytoskeleton.

Enzymes that regulate protein phosphorylation play a crucial role in regulating the cell cycle. **Protein kinases** transfer phosphate groups from ATP to a particular amino acid in the target protein, whereas **protein phosphatases** remove the phosphate groups from the target proteins. The addition and removal of phosphate groups can dramatically alter the behavior of the target proteins. Many kinases and phosphatases show remarkable specificity for the proteins they modify and act as molecular switches that control the activity of their target proteins. Changes in protein phosphorylation are by far the most common examples of the post-translational modifications that regulate the activity of proteins during the cell cycle.

Despite their complexity, most cells are very small. A typical bacterial cell is less than 2 μm long, a typical human cell less than 20 μm in diameter (1 μm [micrometer or micron] is equal to 10^{-6} m). There are, however, some notable exceptions (Figure 1-2). A frog egg, 1 mm in diameter, is large enough to contain about 10^9 bacteria or 10^5 postembryonic frog cells. In a given organism, different types of cells can be of very different sizes, but within a particular cell type, cell size is fairly uniform. We do not understand what governs the size of particular cells.

FIGURE 1-2 Cell size. A frog egg, sea urchin egg, human white blood cell, fission yeast, budding yeast, and bacterium are shown at different magnifications to emphasize the enormous range of cell size. The frog egg is 1 mm in diameter; the bacterium is only 1 μm in diameter.

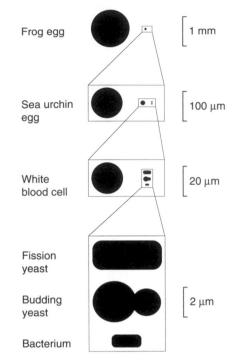

Frog egg — 1 mm

Sea urchin egg — 100 μm

White blood cell — 20 μm

Fission yeast

Budding yeast — 2 μm

Bacterium

Organisms are composed of cells

Although there are many cell types, almost all of them have the ability to grow and divide. Some cells exist as autonomous **unicellular organisms.** Each time such a cell divides, the organism also reproduces, and these cells physically interact only during the sexual phase of the life cycle. Other cells interact and cooperate with each other to form **multicellular organisms.** Most prokaryotes are unicellular, whereas eukaryotes have a wide range of life-styles. Some, like yeasts, are unicellular; others, including most plants and animals, are multicellular.

Multicellular organisms can be thought of as cooperatives devoted to a common end, the growth and reproduction of the organism. The cells grow and divide in complex architectural patterns, differentiate into cells specialized for different functions, form organs, and communicate with each other by chemical and electrical signals. This book discusses the cell cycles of unicellular organisms and multicellular animals. Although the spatial pattern of cell division is exquisitely regulated in plants, we know little about the molecules that regulate plant cell cycles.

In complex animals different cells are specialized for different functions; mammals are estimated to have as many as **200** cell types. Despite their complicated patterns of interdependence, cells in multicellular organisms are distinct entities with considerable autonomy. They are responsible for their own maintenance, and in principle, they can grow and divide independently of their neighbors. In practice, however, organisms regulate cell growth and division according to a set of strict "social" rules. Violation of these rules leads to cancer and ultimately the death of the organism (see Chapter 9).

BASIC PROPERTIES OF THE CELL CYCLE

Earlier, we said that to complete a cell cycle, most cells must replicate and segregate their chromosomes, grow, and divide. A significant minority of cell cycles fail to conform to this simple definition. Rather than attempting a universal definition of the cell cycle, we will simply describe the cycle of a typical human cell and then discuss some of the variations exhibited by other cells.

The cell cycle is divided into four discrete phases

Rapidly dividing human cells have a cell cycle that lasts about 24 hours. The cell cycle is divided into two fundamental parts: **interphase,** which occupies the majority of the cell cycle, and **mitosis,** which lasts about 30 minutes, ending with the division of the cell. During interphase the DNA is diffusely distributed within the nucleus, and individual chromosomes cannot be distinguished. Little activity can be detected with a microscope, although two important classes of processes are occurring. **Continuous processes** occur throughout interphase and are referred to collectively as growth. These processes include the synthesis of new ribosomes, membranes, mitochondria, endoplasmic reticulum, and most cellular proteins. **Stepwise processes** occur once per cell cycle. For example, chromosome replication is restricted to a specific part of interphase called **S phase** (for DNA synthesis). S phase occurs in the middle of interphase, preceded by a gap called **G1** and followed by a gap called **G2** (Figures 1-3 and 1-4). After each chromosome has replicated, the two daughter chromosomes remain attached to each other at multiple points along their length and are referred to as **sister chromatids.** In a typical animal cell cycle, G1 lasts 12 hours, S phase 6 hours, G2 6 hours, and mitosis about 30 minutes.

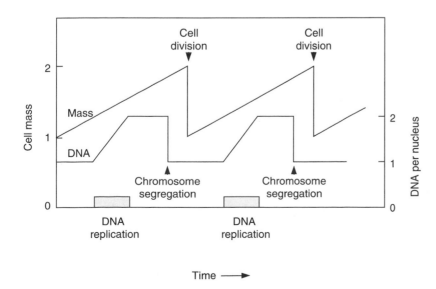

FIGURE 1-3 Cell growth and DNA content during the cell cycle. The variation in the mass of a cell and its DNA content through two cell cycles is shown. Mass increases continuously throughout the cycle. By contrast, DNA content is constant through most of the cycle, increasing during S phase as the DNA replicates, and then falling dramatically during chromosome segregation.

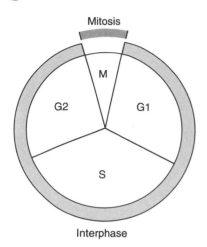

FIGURE 1-4 Stages of the cell cycle. In adult vertebrates the most rapid cell cycles last about one day. Mitosis (M) represents about 5% of the cell cycle, and there is a substantial G1 gap between mitosis and DNA synthesis (S), as well as a G2 gap between replication and mitosis.

Cells assemble a spindle and segregate their chromosomes during mitosis

Mitosis is a dramatic, coordinated change in the architecture of the cell that segregates the replicated chromosomes into two identical sets and initiates cell division (Figure 1-5). The first sign that a cell is about to enter mitosis is a period called **prophase.** During prophase the chromosomes condense; they become visibly distinct from each other as very thin threads that shorten and thicken until each chromosome can be resolved into a pair of sister chromatids.

Prophase ends when the nuclear envelope breaks down, abolishing the distinction between nucleus and cytoplasm. Dramatic changes also occur in the organization of **microtubules,** long fibers that radiate throughout the cell from a **microtubule organizing center** that is called the **centrosome** in animal cells and the **spindle pole body** in yeasts and other fungi. Cells are born with a single microtubule organizing center, which duplicates during interphase. As mitosis progresses, the microtubule network undergoes profound changes that lead to the formation of the **mitotic spindle,** an array of microtubules shaped like an American football with the centrosomes at its ends. (During mitosis, microtubule organizing centers are often called **spindle poles.**) Specialized regions of the chromosomes called **kinetochores** attach them to microtubules. The two members of each pair of sister chromatids attach to microtubules originating from opposite poles of the spindle and move to a position midway between the two spindle poles. When all the chromosomes are lined up like this, the cell is in **metaphase.**

The cell remains briefly in metaphase before dissolving the linkage between the sister chromatids, a critical event that marks the beginning of **anaphase.** Because sister chromatids are no longer held together, they separate from one another and move along the microtubules to opposite poles of the spindle. As the chromosomes near the poles, the physical process of cell division, called **cytokinesis,** begins. A contractile ring pinches the cell into two daughter cells, each containing a complete set of chromosomes and a spindle pole. As cytokinesis proceeds, the chromosomes decondense and acquire a nuclear envelope, re-forming an interphase nucleus, and the microtubule array returns to its interphase pattern.

The relative timing of cytokinesis and chromosome decondensation varies in different cell types, making it difficult to define a precise boundary between the end of mitosis and the beginning of interphase. We consider the onset of anaphase to mark the end of mitosis for two reasons: It is an easily observed and sharply defined event, and it corresponds to an important change in the machinery that regulates the cell cycle. We also choose this point to define the end of one cell cycle and the beginning of the next.

The length of the cell cycle and the details of mitosis vary among different cell types

The cell cycles of different organisms, and even of different cells within the same organism, can appear quite different from each other. For example, the behavior of the nuclear envelope varies between multicellular and unicellu-

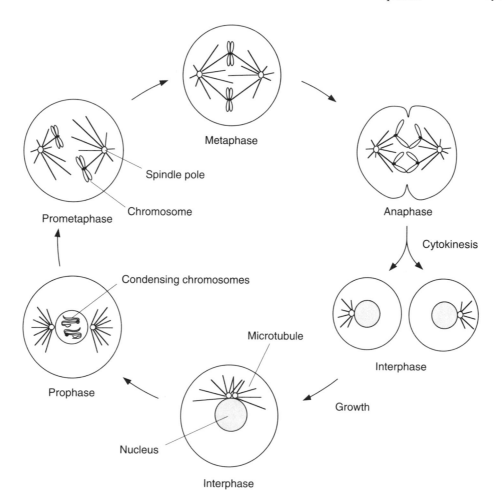

Metaphase

Spindle pole

Chromosome

Prometaphase

Anaphase

Cytokinesis

Condensing chromosomes

Microtubule

Interphase

Prophase

Growth

Nucleus

Interphase

lar organisms (Figure 1-6). Most multicellular organisms have the **open mitosis** just described, in which the nuclear envelope breaks down and then re-forms. In contrast, many unicellular eukaryotes have **closed mitosis,** in which the nucleus stays intact throughout mitosis. In closed mitosis the spindle assembles inside the nucleus, and the nucleus pinches in two after chromosome segregation is complete. The details of cell division also vary among cell types: Animal cells pinch in half, whereas plants and other organisms that have rigid cell walls lay down a new cell wall in the center of the cell. Nevertheless, all these cell cycles retain the four features of the typical eukaryotic **somatic cell cycle:** growth, DNA replication, mitosis, and cell division.

Under appropriate conditions many somatic cells grow continuously, with a more or less constant cell cycle duration, varying from 24 hours for rapidly dividing mammalian cells to 90 minutes for yeast. The duration of the cell cycle varies considerably within a population of cells, however, with one cell often dividing a good deal earlier or later than its neighbors (Figure 1-7). Some cells exist for days, months, or years without growing or dividing but can be induced to divide again. For example, liver cells normally neither grow nor divide, but liver damage rapidly induces them to divide. Other cells, such as neurons, have left the cell cycle irreversibly and can never divide again.

FIGURE 1-5 Stages of mitosis. During prophase the centrosomes move to opposite sides of the nucleus, and the chromosomes condense. The nucleus breaks down, and as the cell progresses to metaphase, the chromosomes align on the center of the spindle. At anaphase the linkage between sister chromatids is broken, and the sisters segregate to opposite poles. Finally, cytokinesis generates two daughter cells.

FIGURE 1-6 Open and closed mitosis. Most multicellular eukaryotes have open mitosis, in which the nuclear envelope breaks down and the chromosomes visibly condense. Many unicellular organisms, such as the budding yeast pictured here (bottom), have closed mitosis: The nuclear envelope does not break down, and chromosome condensation is hard to see. Mitotic chromosome condensation does occur in budding yeast, but it can be seen only with an electron microscope.

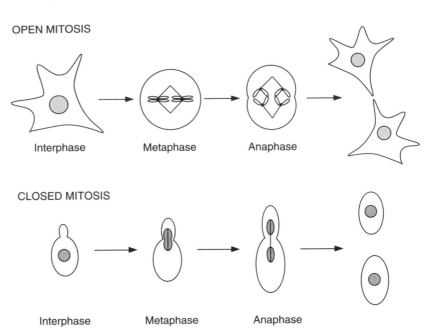

OPEN MITOSIS

Interphase   Metaphase   Anaphase

CLOSED MITOSIS

Interphase   Metaphase   Anaphase

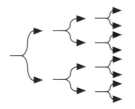

EMBRYONIC CELL CYCLES

Short cycles of constant duration

FIGURE 1-7 Different cell cycles. The shortest cell cycles occur in early embryos and can last as little as 8 minutes. The length of each cycle is very constant. The cell cycle of growing eukaryotic cells lasts from 90 minutes to more than 24 hours, its duration varying considerably within a population of cells. Finally, postembryonic cells can leave the cell cycle and remain for days, weeks, or even years without growing or progressing through the cell cycle.

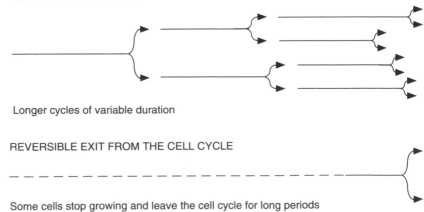

GROWING CELL CYCLES

Longer cycles of variable duration

REVERSIBLE EXIT FROM THE CELL CYCLE

Some cells stop growing and leave the cell cycle for long periods

Cell cycles of early embryos are naturally synchronous

Biochemical studies of the cell cycle require populations of cells that proceed synchronously through the cell cycle. Cells in the somatic cell cycle are usually not synchronous, making it necessary to devise techniques to produce populations that are all in the same part of the cell cycle. There are two ways to synchronize cells. **Induction synchronization** uses special conditions, such as treatment with drugs that block specific steps in the cell cycle, to arrest all the cells at a particular point in the cycle. Removing the block allows the arrested cells to proceed relatively synchronously through the cell cycle. **Selection synchronization** isolates a subpopulation of cells that are at the same point in the cell cycle by exploiting a cellular property that changes as cells go through the cell cycle. One popular method exploits the variation in cell size during the cell cycle by using special centrifuges to sort cells by size. All synchronization methods suffer from a serious limitation. Because the duration of somatic cell cycles varies, the synchrony in a population of cells decays rapidly as they are allowed to pass through the cell cycle.

This disadvantage can be overcome by exploiting the natural synchrony of the **early embryonic cell cycles** that occur in fertilized eggs. It is easy to obtain large quantities of unfertilized eggs from frogs and marine invertebrates and then fertilize them to produce populations of cells that proceed rapidly and synchronously through several cell cycles (see Figure 1-7). Much of our information about the control of the cell cycle has come from studying these highly simplified cycles, which simply subdivide the egg into smaller and smaller cells without any growth between cell divisions (Figure 1-8).

Although rapid, early embryonic cell cycles do obey the rule that rounds of DNA replication alternate with rounds of chromosome segregation. Certain specialized cell cycles, however, break this rule. For example, the meiotic cell cycles that produce the specialized cells involved in sexual reproduction follow one round of DNA replication by not one but two rounds of chromosome segregation (see Chapter 10).

FIGURE 1-8 Growth in somatic and embryonic cell cycles. This idealized representation compares the growth of a population of somatic cells with the cleavage of a fertilized egg. The population of somatic cells maintains a constant cell size because the cells grow after each division. By contrast, the embryo divides without growing, producing smaller cells with each succeeding cell cycle. The embryonic cell cycles are about 20 times faster than those of the somatic cells.

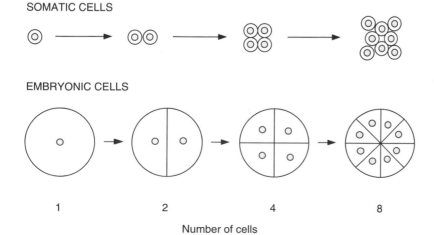

SOMATIC CELLS

EMBRYONIC CELLS

1   2   4   8

Number of cells

APPROACHES TO THE CELL CYCLE

To study the cell cycle we need ways of telling where individual cells are in the cycle. Animal cells in mitosis are easily recognized under the microscope; they lose the distinction between nucleus and cytoplasm, and the condensed chromosomes align on the spindle. Microscopy does not reveal, however, whether an interphase cell is in G1, S phase, or G2.

The DNA content of a cell reveals its position in the cell cycle

Cells in S phase can be detected by their ability to incorporate labeled DNA precursors whose presence can be detected after killing the cells. This method can be modified to determine the fraction of cells in different parts of the cell cycle, but the experiments are time-consuming and indirect. At present most researchers use a technique, called **flow cytometry,** that measures the amount of DNA in each cell of a population. A population of cells is treated with a fluorescent DNA-binding dye and passed one at a time through an exquisitely sensitive device that measures the amount of dye bound to each cell and calculates the amount of DNA present. The DNA content of G2 cells is exactly twice that of G1 cells, and cells in S phase have an intermediate DNA content that increases with the extent of replication. A typical profile of DNA content in an asynchronous, cycling population of mammalian cells is shown in Figure 1-9. There are clear peaks of G1 and G2 cells, but because the cells in S phase are spread over a range of DNA contents, the valley between the G1 and G2 peaks is quite low.

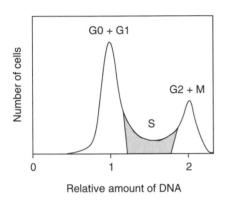

FIGURE 1-9 DNA content varies during the cell cycle. Very sensitive spectrophotometers can measure the amount of DNA in individual cells by detecting the fluorescence from DNA-binding dyes. After many cells are passed through it, the machine (often called a FACS [fluorescence-activated cell sorter] machine) provides a readout of the distribution of cells with different DNA contents in the population. Cells in G2 have twice as much DNA as do cells in G1, and cells in S phase (shaded) have an intermediate amount. The relative heights of the different peaks on the readout reveal the fractions of the population in different phases of the cell cycle.

Cell fusion experiments reveal the logic of the cell cycle

The production of two viable, genetically identical daughter cells poses a number of problems. Cells must replicate their chromosomes accurately and then segregate them into two identical sets. These two activities must be coordinated with each other. For example, since chromosome condensation prevents further DNA replication, cells must not begin mitosis until they have finished DNA replication. This is an example of the **completion problem,** the requirement that certain processes in the cell cycle be finished before others begin. Another problem that cells must solve is the **alternation problem:** What mechanisms ensure that each round of DNA replication is followed by a round of chromosome segregation, rather than another round of DNA replication?

An ingenious series of experiments performed in 1970 revealed the existence of these problems and offered the first clues to how cells solve them. New methods for synchronizing cells made it possible to fuse populations of cells in different parts of the cell cycle to each other. The first set of experiments fused cells in interphase with cells arrested in mitosis. The results were striking (Figure 1-10). Fusing cells in G1, S phase, or G2 to mitotic cells produced fusion products, or **heterokaryons,** in which the interphase nuclei broke down and all the chromosomes condensed. These results suggested that mitosis is dominant to other states in the cell cycle and that mitotic cells contain factors that induce mitosis when introduced into cells in other parts of the cycle.

In the second experiment, S phase cells were fused to G1 cells, after which the G1 nuclei immediately began to replicate their DNA. This result suggested that the cytoplasm of S phase cells contains factors that can induce DNA replication in a G1 nucleus. The heterokaryons made by fusing G1 cells to S phase cells did not enter mitosis until the G1 nucleus had finished DNA replication. This delay revealed the existence of **feedback controls,** cellular

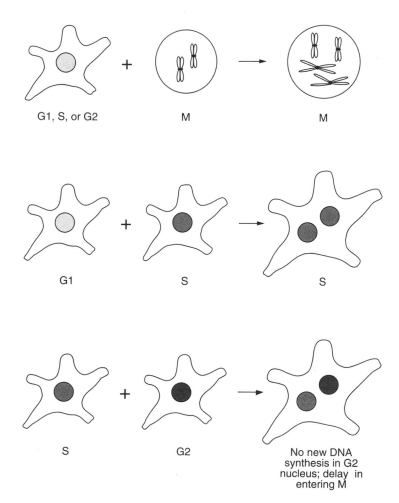

G1, S, or G2   M   M

G1   S   S

S   G2   No new DNA synthesis in G2 nucleus; delay in entering M

FIGURE 1-10 Cell fusion experiments. Fusing mammalian cells at different stages of the cell cycle revealed the logic of the cycle. Fusing cells at any stage of the cell cycle to mitotic cells produced hybrid cells in which all the nuclei entered mitosis (M). Fusing G1 cells to S phase cells induced the G1 nucleus to enter DNA synthesis. By contrast, fusing G2 cells to S phase cells did not induce the G2 nucleus to replicate its DNA, demonstrating a block to rereplication. References: Rao, P. N., & Johnson, R. T. *Nature* **225,** 159–164 (1970); Johnson, R. T., & Rao, P. N. *Nature* **226,** 717–722 (1970).

mechanisms that can arrest the cell cycle at a **checkpoint** until processes in the cell cycle are complete.

Fusing G2 cells to S phase cells produced a surprisingly different result. Although the S phase nuclei continued to replicate their DNA, the G2 nuclei did not initiate a new round of DNA synthesis, and the heterokaryons did not enter mitosis until the S phase cells had finished DNA replication. Since fusion of G1 cells to S phase cells had shown that S phase cytoplasm contained an activator of DNA synthesis, the failure of the G2 cells to indulge in a second round of DNA replication was hard to explain. A simple interpretation of this failure, which also suggests how cells solve the alternation problem, invokes the concept of a **block to rereplication.** This hypothesis postulates that once DNA has been replicated, it becomes modified in some way that makes it incapable of replicating again. If passage through mitosis removes this block, the ability of G1 nuclei to replicate their DNA is neatly explained, as is the necessity for a round of mitosis between each round of DNA replication. The cell fusion experiments represented an important conceptual advance in our understanding of the cell cycle. Unfortunately, there was no easy way to convert these experiments into assays for any of the molecules, such as a mitosis-promoting factor, that they suggested ought to exist.

Two other important types of coordination that occur in the cell cycle were not revealed by the cell fusion experiments. One is the coordination between the stepwise events of the cell cycle and the continuous process of cell growth. Chapters 3 and 4 discuss how the induction of DNA replication and mitosis is regulated by requiring cells to reach a minimum size before they can perform crucial stepwise events in the cell cycle. The other kind of coordination is the regulation of the cell cycle by signals in the cell's external environment, including the presence of nutrients and signals from other cells. How these signals regulate the growth and division of yeast and mammalian cells is discussed in Chapter 6.

CONCLUSION

In this chapter we have seen that the fundamental properties of cells and the key questions about the cell cycle were well defined by 1970. Yet the number of scientists investigating the cell cycle remained conspicuously small until the late 1980s. In the intervening years, understanding the cell cycle was widely recognized as a central problem in biology, but many thought it could not be solved until the basic processes of the cell cycle, such as DNA replication and chromosome segregation, were well understood. Although this attitude seemed highly reasonable at the time, we shall see that, in reality, the conceptual advances in understanding the regulation of the cell cycle did not require a detailed knowledge of chromosome replication or segregation.

SELECTED READINGS

General

Alberts, B., et al. *Molecular biology of the cell,* 2nd ed. 1218 pp. (Garland, New York, 1989). *A comprehensive and detailed text that provides a molecular description of cells and the activities they perform.*

Original Articles

Rao, P. N., and Johnson, R. T. Mammalian cell fusion studies on the regulation of DNA synthesis and mitosis. *Nature* **225,** 159–164 (1970).

Johnson, R. T., and Rao, P. N. Mammalian cell fusion: Induction of premature chromosome condensation in interphase nuclei. *Nature* **226,** 717–722 (1970). *Two papers that describe the result of fusing cells in different parts of the cell cycle.*

THE EARLY EMBRYONIC CELL CYCLE

*M*ANY BIOLOGICAL processes are best studied in cells that are specialized for particular tasks. For example, studies on nerve cells have been the principal route to understanding the electrical properties of cells, and studies on muscle provided the foundation of our knowledge about how chemical energy is converted into movement. Because eggs are specialized for rapid cell division, it is not surprising that studies of them have made a fundamental contribution to unraveling the mysteries of the cell cycle.

Nineteenth-century biologists first appreciated the advantages of eggs for studying the cell cycle, and they meticulously observed and beautifully described the early embryonic cell cycles of marine invertebrate and amphibian eggs. They recognized the division of the cell cycle into interphase and mitosis, and their observations on mitosis compare favorably to modern photographs (Figure 2-1). But because there was no way for them to appreciate (much less understand) the biochemical events that occurred in interphase, their discussions concentrated on the mechanics of mitosis and rarely speculated on the cell cycle as a whole.

In the first half of the twentieth century, interest in eggs as experimental systems gradually declined, reflecting the inability to extend observations from microscopy to biochemistry. During the 1960s, however, progress in biochemical analysis and reproductive physiology, together with the inven-

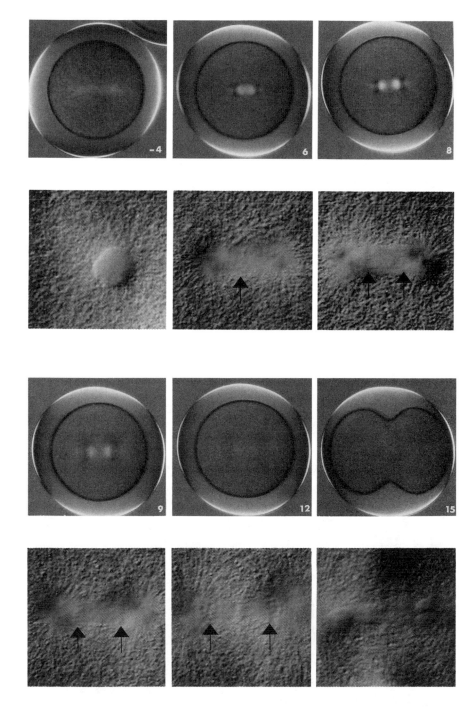

FIGURE 2-1 First mitosis of a fertilized sea urchin egg: a series of micrographs made of living eggs that show the events from the end of the first interphase to the beginning of the second interphase. The numbers are the time in minutes from nuclear envelope breakdown. At each time two views of the egg are shown. Those that show the whole egg within its fertilization envelope are taken by polarization microscopy, which detects highly polarized structures as either brighter or darker than the background (depending on their orientation). The principle structures that this technique detects inside cells are microtubules. The higher-magnification views are taken by differential interference contrast microscopy, which detects gradients in the refractive index of the sample. In these views the positions of the chromosomes are indicated by arrows. Before nuclear envelope breakdown (−4) the nucleus is clearly visible in a sea of yolk granules and other cytoplasmic particles. At metaphase (6) the high density of the spindle microtubules has eliminated most particles from the spindle and the chromosomes are aligned equidistant from the spindle poles. In early anaphase (8) the two sets of chromosomes can be seen separating from each other, and they continue to move rapidly in late anaphase (9). After this the chromosomes start to decondense (12), forming spherical structures that are enclosed by nuclear envelope. Finally, the decondensing chromosomes fuse with each other to form two interphase nuclei (15) as the cell is cleaving into two. Photographs courtesy of John Robinson and Ted Salmon.

tion of techniques to inject materials into living cells, led to a resurgence in the use of eggs for studying the cell cycle. Since then, experiments on eggs have contributed enormously to our knowledge of the cell cycle. Initially the techniques for injecting materials were useful only with very large cells, making frog eggs excellent subjects for analyzing the early embryonic cell cycle. The most widely used source of eggs is the South African clawed frog, *Xenopus laevis.* This chapter describes studies on frog eggs that identified the major inducer of mitosis and led to the idea that a simple biochemical oscillator drives the cell cycle.

EMBRYONIC CELL CYCLES

The eggs of amphibians, marine invertebrates, and insects are large cells that can divide very rapidly. A female frog can produce several thousand eggs, each 1 mm in diameter, that can be fertilized in vitro to produce a large population of cells that proceed synchronously through several cell cycles. The first cell cycle lasts 75 minutes and is followed by 11 synchronous cell cycles, each 30 minutes long (Figure 2-2). These divisions convert the originally solid egg into a hollow ball of cells, called the **blastula,** which then undergoes complex changes in shape that produce a recognizable embryo (Figure 2-3). In early embryonic cell cycles, G1 and G2 are largely suppressed, and the cycle consists essentially of a regular alternation of extremely

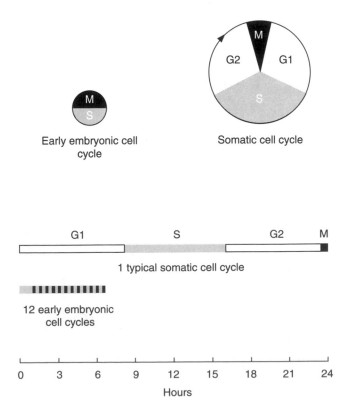

FIGURE 2-2 Early embryonic and somatic cell cycles in frogs. Somatic cell cycles in frogs typically last about 1 day and have a G1 and a G2. Twelve rapid divisions occur after fertilization, after which the synchrony between the cell cycles of neighboring cells breaks down.

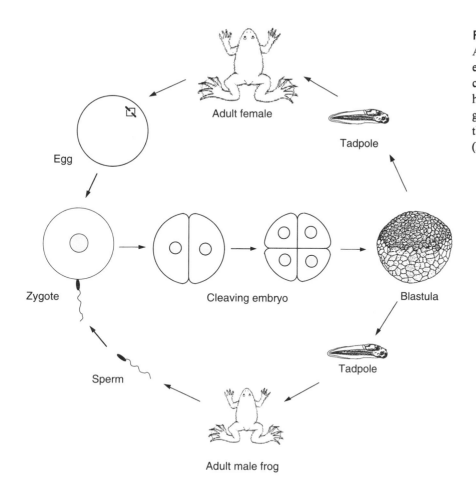

FIGURE 2-3 Frog early development. As they emerge from the female frog, eggs are fertilized and undergo 12 rapid cell divisions to produce the blastula, a hollow ball of cells. The blastula undergoes complex structural rearrangements that ultimately give rise to a tadpole. (Not to scale.)

rapid DNA replication and mitosis. After the twelfth mitosis the cell cycle slows, and the synchrony between neighboring cells breaks down. In this book "early embryonic cell cycle" refers only to these early, rapid, synchronous cell cycles.

The ability of eggs to divide without growing explains why early embryonic cell cycles are faster than somatic cell cycles. Somatic cells are born small and must import nutrients so that they can grow and duplicate all the components of the cell. Eggs, however, are large and inherit from their mothers a stock of nutrients, all the structural components of the cell, and almost all the enzymes that catalyze the processes of the cell cycle. As a result, apart from replicating their DNA, early embryonic cells have to produce only a handful of new components to proceed through a cell cycle that has been stripped to its bare essentials.

Progesterone induces frog oocytes to enter meiosis and become unfertilized eggs

The cells that give rise to eggs are called **oocytes** and begin life at the same size as typical somatic cells. Shortly after birth, oocytes replicate their DNA and then arrest in G2 for about 8 months as the cells grow to a diameter of 1 mm

FIGURE 2-4 Frog oogenesis, meiosis, and fertilization. Oocytes are born at the same size as typical somatic cells. They replicate their DNA and then arrest in G2 as they grow in diameter from 20 μm to 1 mm. Secretion of progesterone by the follicle cells that surround the oocyte induces it to undergo the first meiotic division and enter the second. The oocytes arrest in metaphase of meiosis II. Fertilization overcomes the metaphase arrest and initiates the early embryonic cell cycles.

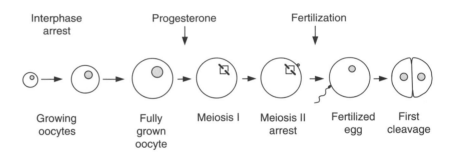

and stockpile the materials needed for the early embryonic cell cycles. To give rise to an egg, the fully grown oocyte must halve its chromosome number to convert itself from a diploid to a haploid cell. The oocyte produces a haploid egg by going through the **meiotic cell cycle,** in which two rounds of chromosome segregation follow a single round of DNA replication (see Chapter 10). At fertilization, the fusion of an egg and a sperm produces a diploid embryo.

When female frogs are appropriately stimulated, the small cells that surround the oocyte secrete progesterone. This hormone acts on the oocyte, leading to nuclear envelope breakdown, chromosome condensation, and the assembly of the meiosis I spindle (Figure 2-4). The process takes about 5 hours, culminating in chromosome segregation and a highly asymmetrical cell division that expels half the chromosomes into a small cell known as the **first polar body.** The oocyte immediately enters meiosis II but arrests in metaphase for several hours as it travels through the oviduct of the frog to emerge as an unfertilized egg. Fertilization releases eggs from their metaphase arrest, allowing them to pass through anaphase of meiosis II, produce a **second polar body,** and enter interphase of the first mitotic cell cycle. The events that convert fully grown oocytes into unfertilized eggs are known as **meiotic maturation** and can be studied in vitro by adding progesterone to oocytes that have been surgically removed from female frogs. Although several important differences exist between the meiotic and mitotic cell cycles (see Chapter 10), studies on the meiotic cell cycle were what first identified the key factor that controls both the meiotic and mitotic cell cycles.

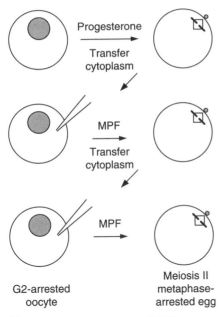

FIGURE 2-5 Discovery of MPF. Oocytes induced to mature into unfertilized eggs by treatment with progesterone are used to donate cytoplasm to untreated oocytes. The transferred cytoplasm contains active MPF which induces the recipient oocytes to enter meiosis. Maturation induces the activation of MPF in the recipient oocyte, allowing it to act as a cytoplasmic donor that can induce meiosis in a fresh round of recipient oocytes. Reference: Masui, Y., & Markert, C. L. *J. Exp. Zool.* **177,** 129–145 (1971).

Cytoplasmic transfer experiments reveal a maturation promoting factor (MPF) that induces meiosis

The mammalian cell fusion experiments described in Chapter 1 showed that mitosis dominates other states in the cell cycle. A similar result was obtained in frog experiments, in which the transfer of cytoplasm from eggs to oocytes demonstrated that meiosis is dominant to interphase. Isolated oocytes were treated with progesterone to induce them to mature into unfertilized eggs. A hollow microneedle was used to remove about 5% of the cytoplasm from an egg and inject it into an oocyte. The recipient oocyte matured as if it had been treated with progesterone (Figure 2-5). The injected oocytes that had matured were then used as cytoplasmic donors to inject fresh oocytes. This second round of recipients also matured, and they could in turn act as donors of cytoplasm that induced yet another round of recipient oocytes to mature. The ability to perform many such successive transfers showed that whatever promotes maturation is a component of the unfertilized egg, rather than progesterone carried over from the initial hormonally matured oocyte. For obvious reasons, the activity that induces maturation was called **maturation promoting factor (MPF).** MPF can also stand for mitosis and meiosis promoting factor, a name that more generally describes its role in the cell cycle.

Progesterone induces MPF activation and nuclear envelope breakdown only if the treated oocytes are allowed to synthesize proteins. Injecting MPF-containing cytoplasm, however, induces maturation even if protein synthesis in the recipient oocyte is inhibited. This observation implies that oocytes must contain molecules, called **preMPF**, that can be converted into active MPF by a series of post-translational reactions.

MPF activation is a purely cytoplasmic process, since oocytes whose nuclei have been removed still produce active MPF when treated with progesterone. Thus, even though MPF regulates the fate of the nucleus, the nucleus is not required for the activation of MPF.

In fertilized eggs the activation of MPF requires protein synthesis and induces entry into mitosis

How widespread is the role of MPF in the cell cycle? Oocytes can be induced to mature by injecting cytoplasm from mitotically arrested mammalian cells, suggesting that MPF exists in a wide range of cell types. The discovery that MPF activity rises and falls in the meiotic and mitotic cell cycles of frog eggs strengthened the suggestion that MPF plays a key role in regulating the cell cycle (Figure 2-6). After treating oocytes with progesterone, a 5-hour lag occurs before MPF activity rises rapidly to a peak at metaphase of meiosis I. Activity falls between the two meiotic divisions, rises again as the meiosis II spindle is assembled, and then remains high during the natural cell cycle arrest in metaphase of meiosis II. Fertilization leads rapidly to inactivation of

FIGURE 2-6 MPF fluctuations in meiotic and mitotic cell cycles. Oocytes have low levels of MPF activity. Progesterone induces the activation of MPF, leading to meiosis I. After a brief decline, a second rise in MPF activity induces meiosis II, and the oocytes remain arrested in metaphase of meiosis II with high levels of MPF. This arrest is overcome by fertilization, which leads to a precipitous decline in MPF activity. Interphase of the first mitotic cell cycle lasts about 60 minutes, while that of cycles 2 through 12 last about 15 minutes. Each mitosis lasts about 15 minutes and is initiated by the activation of MPF and terminated by its inactivation. Reference: Gerhart, J., Wu, M., & Kirschner, M. *J. Cell Biol.* **98**, 1247–1255 (1984).

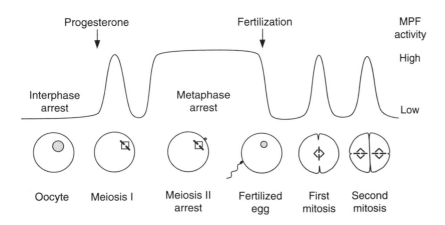

Progesterone   Fertilization   MPF activity

High

Interphase arrest   Metaphase arrest   Low

Oocyte   Meiosis I   Meiosis II arrest   Fertilized egg   First mitosis   Second mitosis

FIGURE 2-7 Induction of mitosis by MPF. Treating fertilized eggs with protein synthesis inhibitors arrests them in interphase, but injecting partially purified MPF into the arrested eggs induces them to enter mitosis. Reference: Newport, J. W., & Kirschner, M. W. *Cell* **37**, 731–42 (1984).

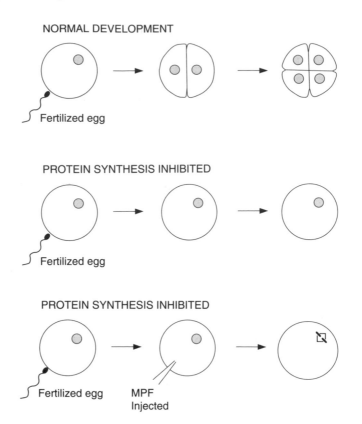

NORMAL DEVELOPMENT

Fertilized egg

PROTEIN SYNTHESIS INHIBITED

Fertilized egg

PROTEIN SYNTHESIS INHIBITED

Fertilized egg   MPF Injected

MPF and interphase of the first mitotic cell cycle. After fertilization, MPF activity rises each time the embryo enters mitosis and then falls as it enters the next interphase.

In early embryonic cell cycles, entry into mitosis and activation of MPF both require protein synthesis in the preceding interphase. The role of MPF in inducing mitosis was investigated by injecting it into embryos that had been arrested in interphase by treatment with protein-synthesis inhibitors. The injected embryos entered mitosis, suggesting that the role of protein synthesis in early embryonic cell cycles is limited to inducing the activation of MPF (Figure 2-7). The presence of maternal stockpiles allows eggs to reduce the role of protein synthesis to controlling the activities that regulate passage through the cell cycle, unlike somatic cells which must synthesize more of each of their components in each cell cycle.

THE CELL CYCLE ENGINE

In most cells, inhibitors of DNA synthesis prevent cells from entering mitosis, and inhibitors of spindle assembly keep them from beginning anaphase, suggesting that the completion of key steps in the replication and segregation of the chromosomes could regulate MPF activity. Surprisingly, experiments on frog eggs led to a very different conclusion.

The cell cycle engine in frog eggs oscillates independently of DNA replication and spindle assembly

When fertilized frog eggs were treated with drugs that inhibit DNA synthesis, the regular rise and fall of MPF activity continued unabated (Figure 2-8). As MPF activity increased, the nuclei broke down, and as it decreased, the nuclei re-formed, even though DNA replication was completely blocked. Treatment with drugs that inhibit spindle assembly also failed to prevent the regular oscillation of MPF activity or the response of the nuclei to its rise and fall. Even in enucleated eggs MPF rose and fell normally. Together these experiments show that, in frog embryos, purely cytoplasmic reactions activate and inactivate MPF and drive the nucleus, whether ready or not, into and out of mitosis. This conclusion clearly violated the dogma derived from studies on somatic cells that the progress of the cell cycle was regulated by the state of the nucleus.

The experiments on frog eggs suggested that the early embryonic frog cell cycle is controlled by a simple biochemical oscillator that periodically drives cells into and out of mitosis. We now know that the oscillator is a series of biochemical reactions in the cytoplasm that collectively lead to the periodic activation and inactivation of MPF. When MPF activity increases, the nuclei respond by breaking down and forming a mitotic spindle. When MPF activity declines, anaphase and cytokinesis follow, and the nucleus reassembles and replicates its DNA. In the early frog embryo, changes in the activity of MPF alter the state of the nucleus, but events in the nucleus do not influence the activation or inactivation of MPF.

Although the independence of the frog egg MPF cycle from the nuclear cycle is unusual, even among embryonic cell cycles, it helps classify events in the cell cycle into two categories: reactions that make up a cell cycle engine and downstream events that the engine controls (Figure 2-9). The **cell cycle engine** consists of all the biochemical components, and the reactions between them, that cause the periodic activation and inactivation of MPF. MPF itself is part of the engine. A major goal of cell cycle research is to enumerate the components of the engine, purify them, and use them to rebuild a working engine in a test tube.

Downstream events lie outside the cell cycle engine and are induced either by active MPF or by the inactivation of MPF. Active MPF induces the downstream events of mitosis, including chromosome condensation, nuclear envelope breakdown, and spindle formation. The inactivation of MPF induces the downstream events that mark the exit from mitosis and the beginning of interphase, including chromosome segregation, chromosome decondensation, nuclear re-formation, and cytokinesis. In early embryonic cell cycles, DNA replication and duplication of the microtubule organizing center are also downstream events induced by the inactivation of MPF. In somatic cell cycles, however, these processes are caused not by the inactivation of MPF but by an additional cell cycle transition, called Start, that occurs during G1, well after the inactivation of MPF (see Chapter 3).

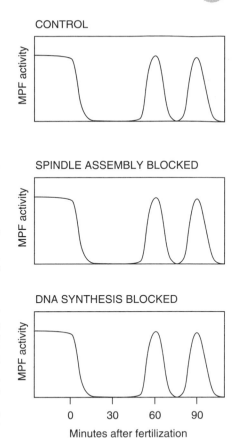

FIGURE 2-8 MPF oscillates independently of DNA synthesis and spindle assembly. The fluctuation of MPF activity in normally fertilized embryos (the control) is compared with that of embryos fertilized in the presence of an inhibitor of DNA polymerization (aphidicolin) or spindle assembly (nocodazole). The oscillations in the three sets of embryos are identical, showing that the failure to complete DNA replication does not influence the reactions that periodically activate and inactivate MPF. Reference: Gerhart, J., Wu, M., & Kirschner, M. *J. Cell Biol.* **98**, 1247–1255 (1984).

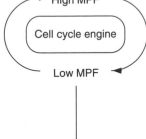

Chromosome condensation
Nuclear envelope breakdown
Spindle assembly

Downstream events of mitosis

High MPF

Cell cycle engine

Low MPF

Downstream events of interphase

Chromosome segregation
Cell division
Chromosome decondensation
Nuclear envelope assembly

DNA replication
Centrosome duplication

FIGURE 2-9 Cell cycle engine and downstream events. The cell cycle is divided into two types of processes. MPF is part of the cell cycle engine, a biochemical machine that produces cyclical oscillations in the activity of MPF and other key cell cycle regulators that control downstream events. Active MPF induces the downstream events of mitosis, which interact to assemble the mitotic spindle. The inactivation of MPF at the end of mitosis induces the downstream events that lead to interphase, including chromosome segregation, and cytokinesis. In early embryonic cell cycles, DNA replication and duplication of the centrosome are consequences of the inactivation of MPF.

Transitions in the cell cycle engine activate and inactivate MPF

Dividing the processes of the cell cycle into a cell cycle engine and a set of downstream events helps to reveal the logic of the cell cycle. The early embryonic cell cycle engine has two states: mitosis, where MPF is active, and interphase, where it is inactive. Transitions between these states induce the downstream events that produce profound rearrangements of the cell. After each rearrangement is complete, the architecture of the cell remains unchanged until the next transition in the engine occurs (Figure 2-10). Active MPF induces the changes that culminate in metaphase, and the cell does not enter anaphase and progress into interphase until the transition that inactivates MPF. Thus, the visible events of the cell cycle reveal transitions in the state of the cell cycle engine. The main reason that we consider the onset of anaphase as marking the end of mitosis is that it occurs very shortly after the inactivation of MPF, so the end of mitosis directly reflects a change in the state of the cell cycle engine.

The extent to which cell cycle engine is controlled varies. In unfertilized frog eggs the cell cycle engine is arrested with high levels of MPF. Fertilization restarts the engine, which then runs freely, whether or not downstream events are completed successfully. In many embryos and all somatic cell cycles, each transition in the engine is regulated in response to the completion of downstream events.

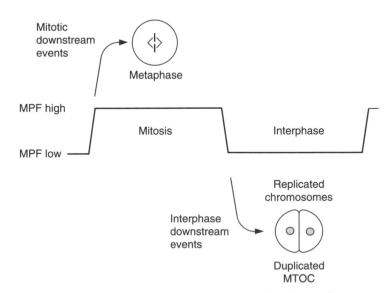

FIGURE 2-10 Organization of the early embryonic cell cycle. Each transition in the cell cycle engine leads to coordinated changes that rearrange the cell. The cell remains in this new state until the next transition occurs in the cell cycle engine. The activation of MPF at the beginning of mitosis leads to the mitotic downstream events, which produce a cell with a metaphase spindle. The inactivation of MPF induces the interphase downstream events, which produce a G2 cell with replicated DNA and duplicated centrosomes (MTOC).

The destruction of cyclins at the end of each mitosis suggests a simple model for the cell cycle

What makes the cell cycle engine oscillate? The experiments on frog eggs suggested that the regular oscillations in MPF activity drive the cell cycle but did not reveal the molecular nature of MPF or the enzymes that turn it on and off. The only clue was that protein synthesis was required to activate MPF in every cell cycle. Perhaps the newly made proteins were components, or activators, of MPF. With the benefit of hindsight, it now seems obvious that in either case the critical proteins would have to be used up or destroyed by passage through mitosis. If this were not true, it would be hard to explain why new protein synthesis was required to activate MPF. Many unsuccessful attempts were made to identify **periodic proteins,** whose periodic synthesis had been proposed to induce cell cycle transitions. Then, as often happens, what had been sought diligently was stumbled upon accidentally.

The breakthrough came from studies of protein synthesis in sea urchin eggs. Newly fertilized eggs were incubated with a radioactive amino acid and sampled every 10 minutes to analyze the pattern of radioactively labeled proteins. The amount of radioactivity in most proteins increased continuously throughout the experiment, but one protein behaved quite differently. It disappeared abruptly at the end of each mitosis and then gradually reappeared during the next interphase (Figure 2-11). Additional experiments showed that **cyclin,** as it is now called, is a periodic protein synthesized throughout the cell cycle but degraded at the end of each mitosis. The assumption that periodic proteins are periodically synthesized, rather than continuously synthesized and periodically destroyed, may explain why cyclin remained unidentified for so long.

The behavior of cyclin suggested a simple model of the cell cycle engine, in which the activation of MPF drives cells into mitosis and the decline of MPF activity leads to the next interphase. The model rests on three postulates: the accumulation of cyclin during interphase activates MPF, active MPF induces the destruction of cyclin, and the destruction of cyclin leads to the inactivation of MPF. In this model the cell cycle engine flip-flops periodically between mitosis and interphase (Figure 2-12). The accumulation of cyclin in interphase leads to the activation of MPF and the induction of mitosis, and the ability of MPF to induce the degradation of cyclin leads to the inactivation of MPF and the next interphase. The model made three simple, testable predictions:

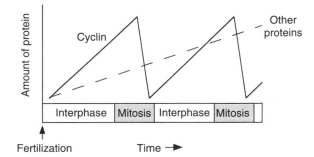

FIGURE 2-11 Discovery of cyclin. The abundance of newly synthesized proteins was measured during the first two cell cycles of fertilized sea urchin eggs. The amount of most new proteins increased continuously after fertilization. The abundance of cyclin, however, increased during each interphase and declined abruptly at the end of each mitosis. Reference: Evans, T., Rosenthal, E. T., Youngblom, J., Distel, D., & Hunt, T. *Cell* **33,** 389–396 (1983).

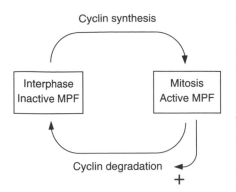

FIGURE 2-12 Early embryonic cell cycle model. The fluctuation of cyclin abundance and MPF activity in the early embryonic cell cycle suggests a flip-flop model for the cell cycle engine. In interphase MPF is inactive, but the accumulation of cyclin leads to the activation of MPF and entry into mitosis. In mitosis MPF is active, but MPF induces the degradation of cyclin, leading to the inactivation of MPF and entry into interphase.

Cyclin is a protein that activates MPF or is a part of MPF, cells must accumulate cyclin to enter mitosis, and cells must degrade cyclin to exit mitosis. Chapter 4 describes experiments that confirmed these predictions and led ultimately to a more complex but more realistic picture of the cell cycle engine.

Cyclins are found in all eukaryotes that have been examined, including yeasts, coelenterates, flies, echinoderms, mollusks, amphibians, mammals, and plants. We now know that there are many different cyclins, which form a large family of related proteins with different functions. Mitotic cyclins (cyclin B) are components of MPF, S phase cyclins (cyclin A) play a poorly defined role in the control of DNA replication, and G1 cyclins are important in catalyzing the events that move the somatic cell cycle from G1 into S phase.

CONCLUSION

The experiments on MPF in frog eggs and oocytes and the mammalian cell fusion experiments (see Chapter 1) both led to the conclusion that mitotic cells contain a dominant inducer of mitosis. In other respects, however, the two sets of experiments led to strikingly different conclusions. The studies of frog eggs revealed a cytoplasmic oscillator that drives the nucleus into mitosis even if DNA replication is unfinished, whereas in the fusion experiments cells entered mitosis only after DNA synthesis was finished. For many years the puzzling contrast between the dictatorship of the cell cycle oscillator in frog eggs and the interdependence of processes in the somatic cell cycle suggested that the early embryonic and somatic cycles were fundamentally different from each other.

The ability to assay material for MPF activity meant that oocyte injection offered a method for characterizing and purifying MPF that the cell fusion experiments did not offer. Nevertheless, two problems hindered research on MPF. The first was purely practical: MPF proved extraordinarily difficult to characterize and purify, partly because the oocyte injection assay was cumbersome and technically demanding. Second, the possibility that MPF was active only in the specialized meiotic cell cycle meant that few people studied it until the evidence for its role in the mitotic cell cycle was overwhelming. As we shall see in Chapter 4, the pace of research on MPF increased exponentially during the 1980s, leading to the discovery that it is a universal component of a highly conserved cell cycle engine found in all eukaryotes.

SELECTED READINGS

General

Wilson, E. B. *The cell in development and heredity,* 3rd ed. 1232 pp. (Macmillan, New York, 1928). *This classical summary of the work of early light microscopists reveals the strengths and weaknesses of a purely descriptive approach to the cell cycle.*

Original Articles

Masui, Y., & Markert, C. L. Cytoplasmic control of nuclear behavior during meiotic maturation of frog oocytes. *J. Exp. Zool.* **177,** 129–145 (1971). *The discovery of MPF.*

Gerhart, J., Wu, M., & Kirschner, M. Cell cycle dynamics of an M-phase-specific cytoplasmic factor in Xenopus

laevis oocytes and eggs. *J. Cell Biol.* **98,** 1247–1255
(1984). *Inhibitors of spindle assembly fail to prevent
the regular oscillation of MPF activity.*

Newport, J. W., & Kirschner, M. W. Regulation of the cell
cycle during early Xenopus development. *Cell* **37,**
731–742 (1984). *Injection of MPF into interphase
cells induces them to enter mitosis.*

Evans, T., Rosenthal, E. T., Youngblom, J., Distel, D., &
Hunt, T. Cyclin: A protein specified by maternal
mRNA in sea urchin eggs that is destroyed at each
cleavage division. *Cell* **33,** 389–396 (1983). *The discovery of cyclin.*

3 GENETIC ANALYSIS OF THE CELL CYCLE

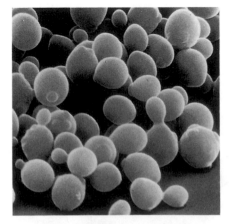

FIGURE 3-1 Budding yeast. This scanning electron micrograph shows a population of budding yeast cells. Micrograph courtesy of Eric Schachtbach and Ira Herskowitz.

*T*HIS CHAPTER describes the genetic analysis of the cell cycle in the budding yeast *Saccharomyces cerevisiae* (Figure 3-1), a unicellular organism that has been used for centuries in the production of bread, beer, and wine. Since the 1940s, the efforts of many scientists have developed techniques for genetically analyzing budding yeast to the point where it is the organism of choice for genetic studies of the cell cycle.

Unlike eggs, growing cells must import nutrients from a changeable environment, rather than relying on resources provided to them by their mothers. How do growing cells coordinate the cell cycle engine with cell growth in a way that produces a population of cells of fairly uniform size? How does the cell cycle engine respond to starvation or the failure to complete a downstream event like DNA replication? Genetic analysis of the cell cycle in yeast has answered these questions, and strategies used by budding yeast are used by most other cells.

Budding yeast introduces us to the somatic cell cycle, whose organization differs from that of the early embryonic cell cycle. In early embryonic cell cycles, the inactivation of MPF and exit from mitosis commit cells to replicating their DNA and microtubule organizing center. In the somatic cell cycle, however, an additional transition, which takes place during G1, must occur before the cell can initiate replication (Figure 3-2). This transition is called **Start** in unicellular eukaryotes and the **restriction point** in multicellular ones, although we believe that it is fundamentally the same in both types of cell.

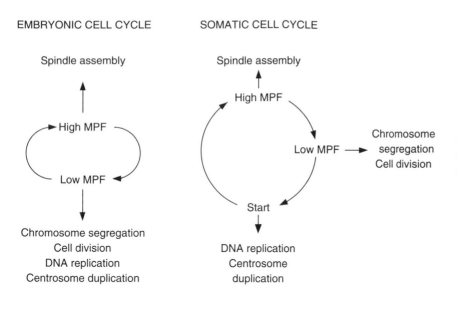

FIGURE 3-2 Organization of early embryonic and somatic cell cycles. The embryonic cell cycle engine has only two transitions, the activation and inactivation of MPF. The inactivation of MPF induces both chromosome segregation and DNA replication. In somatic cell cycles the engine has an extra transition, Start, which must occur before cells can replicate their DNA.

Start divides interphase into two periods. Before Start, cells are in G1; they have unreplicated chromosomes and a microtubule organizing center (Figure 3-3). Passage through Start induces replication of DNA and the microtubule organizing center, leading to a new state that corresponds to G2 in the conventional view of the cell cycle. A major theme of this chapter is that Start acts as a **checkpoint** at which the progress of the cell cycle engine can be regulated by events inside and outside the cell.

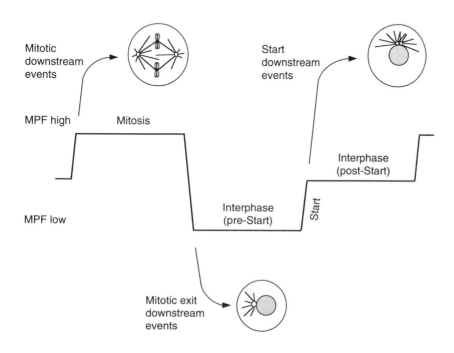

FIGURE 3-3 Organization of the somatic cell cycle. Transitions in the somatic cell cycle engine lead to three different cell states. Activating MPF induces downstream events that produce metaphase, inactivating MPF induces downstream events that produce cells in G1 (with unreplicated DNA and microtubule organizing centers), and passage through Start induces downstream events that produce cells in G2 (which have replicated these components).

CELL DIVISION CYCLE MUTANTS

Genetic analysis of the cell cycle, which began in the late 1960s, was motivated by the phenomenal success of bacterial genetics in explaining the principles that regulate gene expression. In particular, the genetic analysis of bacterial viruses showed how ordered sequences of gene expression control viral replication. In addition, studies on mutations in the structural components of these viruses suggested rules for the assembly of complicated structures built from many different proteins. Although these successes stimulated genetic analysis of the cell cycle, the way in which cells ensure an ordered series of transitions in the cell cycle is rather different from the strategies that control the assembly of bacterial viruses.

Budding yeast is the organism of choice for genetic analysis of the cell cycle

Studying the cell cycle in budding yeast has many advantages. This organism grows rapidly in a simple medium containing glucose and amino acids, dividing every 90 minutes. Its genetics are powerful and fast, making it easy to isolate mutations in basic cellular processes and to clone the genes identified by these mutations. Specific genes can be deleted or overexpressed at will, and within the next few years the DNA sequence of all 16 chromosomes that form the yeast genome will be known.

Because *S. cerevisiae* divides by budding, it is easy to follow the progress of the cell cycle in living cells. Early in the cell cycle a small bud forms, which enlarges continually and ultimately separates from the mother cell (Figure 3-4). Bud formation occurs just after Start, providing a convenient morphological marker for this event. Moreover, the ratio between the size of the mother and the bud gives a rough estimate of a cell's position in the cell cycle.

Budding yeast cells differ from mammalian cells not only in their style of growth, but also in the organization of mitosis. First, as in many unicellular eukaryotes, the nuclear envelope of a budding yeast cell does not break down during mitosis (see Figure 1-6), and microtubules are nucleated by spindle pole bodies rather than by centrosomes (see Chapter 1). Second, in most cells, spindle assembly cannot begin until DNA replication is complete. In budding yeast, however, MPF activation and spindle assembly can occur during S phase and are not delayed by treatments that block replication. Finally, for many years it was impossible to detect mitotic chromosome condensation in budding yeast. Taken together, these observations led to the idea that mitosis in yeast might be fundamentally different from mitosis in other eukaryotes. Recently, however, chromosome condensation has been directly observed by light and electron microscopy. The finding that chromosome condensation

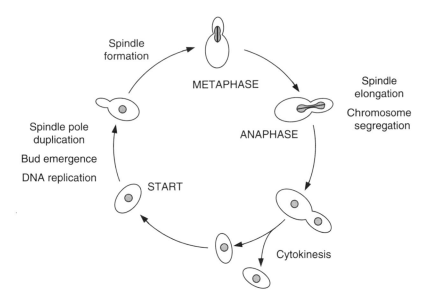

METAPHASE

Spindle
formation

Spindle
elongation

Chromosome
segregation

ANAPHASE

Spindle pole
duplication

Bud emergence

DNA replication

START

Cytokinesis

FIGURE 3-4 Budding yeast cell cycle. In budding yeast, cells can bud only after they have passed Start and can divide only after they have inactivated MPF, making it easy to monitor these cell cycle transitions in living cells. There is, however, no simple morphological or biochemical marker for entry into mitosis.

does not occur until DNA replication is complete strongly suggests that budding yeast has a cell cycle transition that corresponds to entry into mitosis in other cells. Like spindle assembly in other cell types, this transition cannot occur until DNA replication is complete.

Cell division cycle mutants arrest the cell cycle at specific points

The search for mutations that affected the budding yeast cell cycle began in the late 1960s. Because cell division is essential for life, mutations that affect the cell cycle must be isolated as **conditional mutations,** which inactivate the product of the gene under one set of conditions but not another. The most common conditional mutants are **temperature-sensitive *(ts)* mutants,** in which the gene product can function at one temperature, called the **permissive temperature,** but not at a higher, **restrictive** (or nonpermissive) **temperature.** For budding yeast, the permissive temperature is usually room temperature (20–23°C), and the restrictive temperature is 35–37°C.

The first cell cycle mutants came from a large collection of temperature-sensitive mutants that could grow at 23°C but not at 36°C. When asynchronous cultures of most of the temperature-sensitive mutants were shifted to 36°C, the ratio of bud to mother size in the arrested cells was highly variable, indicating that different cells in the population had arrested at different positions in the cell cycle. But a small fraction of the temperature-sensitive mutants showed a remarkable phenotype: After shifting them to 36°C, all the cells in the population arrested with the same morphology, showing that they had accumulated at a particular point in the cell cycle (Figure 3-5). This

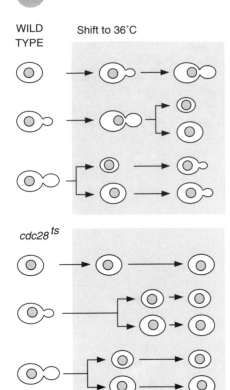

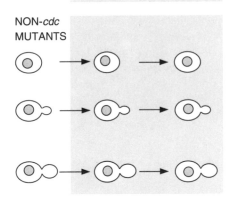

FIGURE 3-5 Cell division cycle mutants. Temperature-sensitive *cdc* mutants are distinguished from other temperature-sensitive mutants by their behavior at the restrictive temperature. Most temperature-sensitive mutations arrest cells at random positions throughout the cell cycle when an asynchronous population of cells is shifted to the restrictive temperature. A *cdc* mutant, however, arrests all cells in the population at the same point in the cell cycle. Reference: Hartwell, L. H., Culotti, J., & Reid, B. *Proc. Natl. Acad. Sci. USA* **66**, 352–359 (1970).

uniform cell cycle arrest suggested that the mutant gene product is required only at a specific point in the cell cycle. Thus, the mutations were named **cell division cycle (cdc) mutants.** The temperature-sensitive cell division cycle (*cdc^{ts}*) mutants were initially classified by two criteria: their morphology when arrested at 36°C and the downstream events whose initiation or completion was blocked at this temperature. The original search identified *cdc^{ts}* mutants in 32 different genes; since then the known number of *CDC^+* genes has increased to over 50 (Appendix).

Interactions between *cdc^{ts}* mutants define a pathway of events in the cell cycle

The analysis of mutations that affect the assembly of bacterial viruses had revealed the logical organization of their assembly pathways. During the 1970s geneticists pursued the idea that a sophisticated analysis of the yeast *cdc^{ts}* mutants would reveal the logical organization of the cell cycle. As part of this approach, they determined which downstream events each *cdc^{ts}* mutant could complete at the restrictive temperature. For example, they found that if a *cdc^{ts}* mutant that blocked DNA replication was shifted up to the restrictive temperature during S phase, the cells arrested in S phase and did not perform nuclear division and cytokinesis. Thus it was clear that mitosis and cell division require the completion of DNA replication. Similarly, *cdc^{ts}* mutants that prevented nuclear division blocked cytokinesis, showing that cytokinesis is also dependent on nuclear division.

Detailed analysis of all the *cdc^{ts}* mutants produced the logical map of the cell cycle shown in Figure 3-6. Start induces three independent pathways, of which one leads to budding, another to spindle pole body duplication, and the third to the initiation of DNA replication. Nuclear division is dependent on successful completion of spindle pole body duplication and DNA replication but does not require successful budding; cytokinesis, however, is dependent both on budding and on the completion of nuclear division. No event requires successful cytokinesis, and under certain nutritional conditions, cytokinesis is suppressed in wild-type strains, producing long chains of interconnected cells.

The genetic analysis of budding yeast *cdc^{ts}* mutants presented a very different picture of the cell cycle from the one derived from studies on frog eggs. At first glance, the complex interdependencies in the yeast cell cycle bore no resemblance to the free-running cell cycle engine of the frog egg. We now know that these apparent differences conceal the fundamental similarity of the cell cycle engine in all eukaryotes (see Chapter 4).

START

The discovery that initiating one process in the cell cycle often depends on completing an earlier process suggested that *cdc^{ts}* mutants could be divided into two classes. One class affects components of the cell cycle engine itself

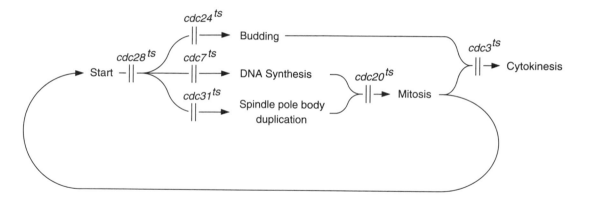

and arrests the cycle because the engine stops at the restrictive temperature. The other class affects components involved in downstream events and arrests the cell cycle engine only because feedback loops make progress of the cycle dependent on the completion of downstream events. Detailed analysis of the *cdc*[ts] mutants that arrest cells early in the cell cycle distinguished these two classes of mutants and defined the cell cycle transition that we now call Start.

The *cdc28*[ts] mutant arrests the cell cycle in G1 and defines Start

In well-fed budding yeast, three events occur early in interphase: The cells bud, DNA synthesis begins, and the spindle pole body duplicates. Some mutants block only one of these three events at the restrictive temperature. For example, *cdc7*[ts] mutants cannot initiate DNA synthesis, *cdc24*[ts] mutants cannot bud, and *cdc31*[ts] mutants cannot duplicate their spindle pole bodies (Figure 3-7). Thus, these three mutations identify genes whose products participate in particular downstream events in the cell cycle.

A fourth mutant, *cdc28*[ts], dramatically altered the prevailing view of the cell cycle and defined one of its key events. At the restrictive temperature, the *cdc28*[ts] mutation blocks all the early downstream events in the cell cycle. Although the cells continue to grow in size, they cannot bud, duplicate their spindle pole bodies, or replicate their DNA. The ability of a single mutation to block all these processes suggested that the Cdc28 protein was a key component of the cell cycle engine. In addition, the discoverers of *cdc28*[ts] argued that it identified Start as a transition in the cell cycle that initiates the processes that lead ultimately to cell division. Start was defined as the point in the cell cycle at which budding, DNA replication, and spindle pole body duplication become insensitive to loss of *CDC28*[+] function. Once cells have passed this point, they are irrevocably committed to replicating their DNA. Although its definition identifies Start as a particular point in time, Start represents the end of a process that commits the cell to replicating its DNA, just as the activation of MPF represents the end of a series of biochemical reactions that commits the cell to assembling a spindle and segregating its chromosomes.

FIGURE 3-6 Logical map of the cell cycle. Analysis of *cdc*[ts] mutants in budding yeast produced this scheme for the interactions of different processes in the yeast cell cycle. The processes blocked by mutants are indicated by the paired vertical bars and the number of the *cdc*[ts] mutant. Note that budding and cytokinesis are not required for progress through the cell cycle, whereas failure of DNA replication, spindle pole body duplication, and spindle assembly all arrest the cycle. Reference: Hartwell, L. H., Culotti, J., Pringle, J. R., & Reid, B. J. *Science* **183**, 46–51 (1974).

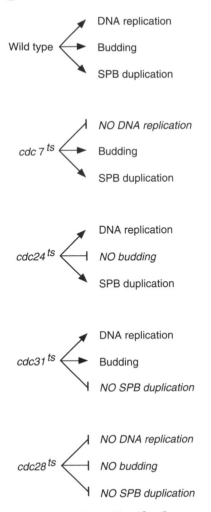

FIGURE 3-7 *cdc28^ts^* identifies Start. Phenotypes of the *cdc7^ts^*, *cdc24^ts^*, and *cdc28^ts^* mutants are shown The *cdc7^ts^*, *cdc24^ts^*, and *cdc31^ts^* mutants block individual downstream events that are dependent on Start. The *cdc28^ts^* mutation allows cell growth to continue but blocks all downstream events dependent on Start, thus identifying an essential protein, whose activity is required to pass Start. Reference: Hartwell, L. H., Culotti, J., Pringle, J. R., & Reid, B. J. *Science* **183**, 46–51 (1974).

Mating factors inhibit passage through Start

Two types of haploid budding yeast cells exist, **a** and α, which can mate with each other to form diploid cells. To mate, two cells must not only be of different mating types, but they must also both be in the part of interphase between the end of mitosis and Start. Once cells have passed through Start, they cannot mate until they enter the next cell cycle. The mammalian cell fusion experiments described in Chapter 1 help to explain why cells must be in the same phase of the cell cycle when they mate. When a mitotic cell is fused to a cell in interphase, the interphase nucleus enters mitosis whether it has finished DNA replication or not. Thus, the mating of a mitotic cell with an interphase cell would inflict severe damage on the genome of the interphase cell.

Once they are arrested, the **a** and α cells mate by forming specialized projections that grow toward each other. When these projections meet, the cell walls dissolve at the point of contact, and the plasma membranes of the two cells fuse to create a common cytoplasm (Figure 3-8). Subsequently, the two haploid nuclei fuse to form a single diploid nucleus, and the newly formed diploid cells pass through Start and reproduce by budding.

Mating partners sense each other's presence by chemical signaling. A small peptide (called **a** factor) secreted from **a** cells is detected by receptors on the surface of α cells, while α cells in turn secrete α factor, a different peptide that binds to receptors on the surface of **a** cells. Cells do not have the receptor for the mating factor that they themselves secrete. When mating factors bind to their receptors, they activate a complex intracellular signaling pathway that has two branches. One arrests the cell cycle at Start; the other induces the expression of genes involved in the physical process of mating (see pages 90–96). The signaling machinery can be activated throughout the cell cycle, but only cells that have completed mitosis and have not yet passed Start are subject to cell cycle arrest and can become competent to conjugate. Thus, cells in S phase that are exposed to mating factors must complete DNA replication and mitosis before they can mate.

The peptide α factor is a useful tool for synchronizing populations of budding yeast. An exponentially growing culture of **a** cells is treated with α factor until all the cells have arrested at Start. After the mating factor is washed out of the culture, the cells proceed synchronously through Start.

Starvation induces diploid budding yeast to sporulate

Although diploid budding yeast cells cannot mate, they can enter an alternative cell cycle if they are starved of nitrogen sources and restricted to a nonfermentable carbon source (such as ethanol). These conditions induce cells to enter the meiotic cell cycle, leading to the formation of four haploid spores that are more resistant than mitotic cells to heat and desiccation (Figure 3-9). On rich medium the spores germinate and give rise to haploid cells.

In nature, two different strategies ensure that these haploid cells rapidly find partners and mate. The first is that each meiotic cell produces two **a** and

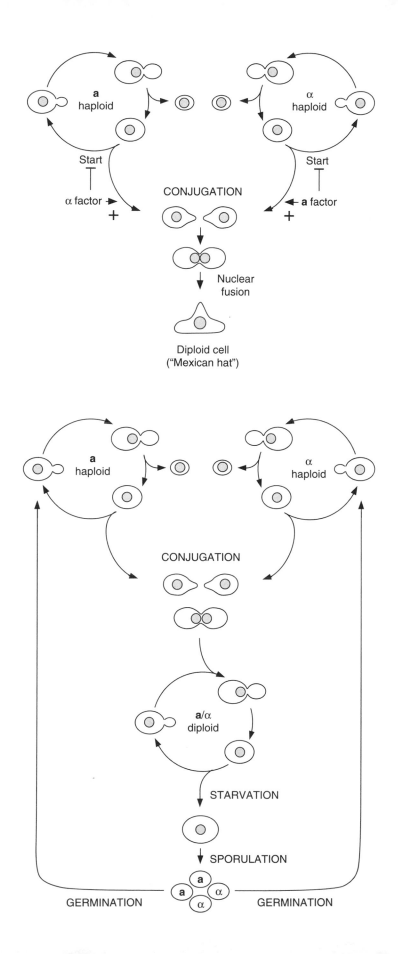

FIGURE 3-8 Mating in budding yeast. Yeast can reproduce asexually by budding, as either haploid or diploid cells. Haploid cells exist in two mating types, **a** and α, and cells of opposite mating types can mate to form diploid cells. Haploid budding cells secrete mating factors that bind to receptors on each other's surface. This chemical signaling induces cell cycle arrest and the expression of genes that carry out the physical process of mating.

FIGURE 3-9 Budding yeast life cycle. If starved of nitrogen and given nonfermentable carbon sources, the diploid cells will sporulate to yield four haploid spores. These spores germinate to form haploid budding cells that can mate with each other to restore the diploid state.

two α cells within its own cell wall, so when the spores germinate, those of opposite mating types have a good chance of mating. The second is that, after germinating and dividing once, haploid cells can switch from one mating type to another, producing a mixed population of **a** and α cells that can mate. This strategy, called **mating type switching** (or homothallism), allows spores that germinate in isolation to give rise to diploid cells. Most laboratory yeast strains carry mutations that prevent mating type switching in order to allow cells to grow stably as haploids.

The behavior of different *cdc*^{ts} mutants was analyzed to determine when cells become committed to the mitotic rather than the meiotic cell cycle (Figure 3-10). The mutants were arrested at the restrictive temperature in rich medium and then transferred to sporulation medium at the permissive temperature. Cells arrested at Start by *cdc28*^{ts} mutations sporulated in the same cell cycle when they were shifted to sporulation medium at the permissive temperature. Cells arrested by mutations in the *cdc4*^{ts} and *cdc7*^{ts} genes, which block the initiation of DNA replication, could also enter the meiotic cell cycle immediately when they were transferred to sporulation medium. Cells arrested by any *cdc*^{ts} mutation that arrested the cell cycle after the initiation of replication were committed to the mitotic cell cycle: After they were returned to the permissive temperature, they had to complete the current cell cycle before entering meiosis.

Similar experiments on haploid cells defined Start as the point in the cell cycle at which cells lose the ability to mate. Comparing the experiments on haploid and diploid cells reveals that the two types of cells lose the ability to mate and sporulate at slightly different points in the cell cycle. As soon as haploid cells pass Start they can no longer mate, but diploid cells do not lose the ability to enter the meiotic cell cycle until they begin DNA replication. Since, except in certain *cdc*^{ts} mutants, DNA replication begins soon after Start, it is hard to assess the significance of this difference, which might simply

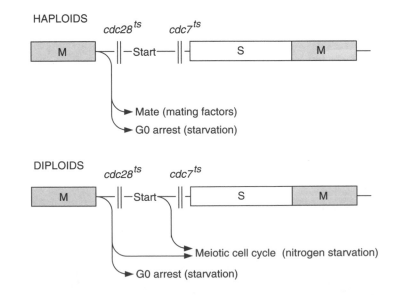

FIGURE 3-10   Commitment to the mitotic cell cycle. The cell cycles of haploid and diploid budding yeast cells are represented as horizontal lines. The stages at which haploid cells can leave the mitotic cell cycle to mate and diploid cells can leave to enter the meiotic cell cycle are indicated.

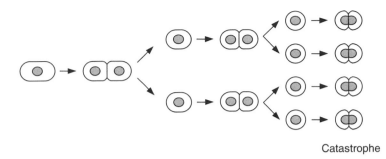

Catastrophe

FIGURE 3-11 Coordinating cell growth and division. Cells that can complete a cell cycle in less time than it takes to double in size become progressively smaller with each generation, until they reach a size at which division kills the cells.

reflect a difference between the cell cycles of haploids and diploids. The important conclusion is that once cells have begun DNA replication, they are committed to the mitotic cell cycle. By regulating DNA replication Start acts as a checkpoint that limits a cell's options to enter alternative cell cycles.

Start coordinates the cell cycle engine with cell growth

In addition to replicating and segregating their chromosomes, almost all cells must grow between each cell division. All known eukaryotic cells can replicate their DNA and segregate their chromosomes faster than they can double in mass. In other words, it is possible to complete a cell cycle in less time than it takes the cell to double in size. Therefore an unrestrained cell cycle could run faster than cell growth, leading to a population of cells that would become progressively smaller with each cell cycle (Figure 3-11).

In theory, cells could avoid this problem if they obeyed the simple rule that they exactly doubled in size before they divided. But Figure 3-12 shows that this solution would fail in budding yeast, which divides asymmetrically to produce a daughter cell that is always smaller than its mother. If the daughter cell doubled in size and divided asymmetrically to produce a granddaughter of

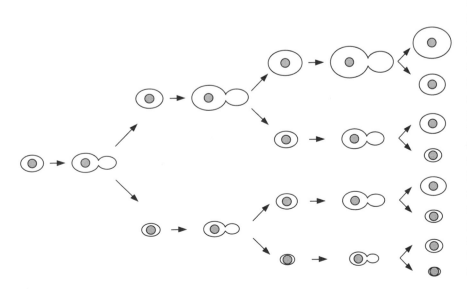

FIGURE 3-12 Problem of unequal cell division. This diagram shows what happens if cells obeying the simple rule of exactly doubling their birth mass before dividing divide unequally to produce one cell that is bigger and one that is smaller than the average birth size. To produce one average-size granddaughter, the small daughter must divide unequally, making the other granddaughter very small. The opposite is true for the large daughter, which can produce a normal-size granddaughter only at the expense of producing a very large granddaughter. Thus, the simple mass-doubling rule soon produces a population that contains very large and very small cells.

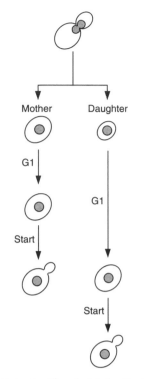

FIGURE 3-13 Threshold size for Start. The daughter cells of budding yeast, which are born smaller than their mothers, must grow more than their mothers to reach the minimum size required to pass Start. The interval between birth and Start is therefore much longer in daughters than it is in mothers. Reference: Hartwell, L. H., & Unger, M. W. *J. Cell. Biol.* **75**, 422–435 (1977).

the original mother cell, the birth mass of the granddaughter would be smaller than that of the cell that gave rise to it. If this process continued for many generations, the great-great-great-granddaughters of the original cell would be vanishingly small. Although this problem is most obvious in cells where the asymmetry of cell division is conspicuous, it applies to all cell types, since there is no mechanism for ensuring that division produces two daughters of exactly the same size.

Analyzing the difference between the cell cycles of mothers and daughters in budding yeast revealed the mechanism that coordinates growth with the cell cycle. The cell cycle of daughter cells is longer than that of mothers, and daughters bud much later than their mothers. Because yeast cells grow throughout the cell cycle, the longer interval between birth and budding means that when daughters bud, they are as big as their mothers were when they gave birth. The late budding of daughters strongly suggested that Start occurs later in the cell cycle of daughters than it does in that of mothers. This prediction was confirmed by using *cdc28^ts* mutants to show that daughters pass Start much later in the cell cycle than their mothers. Altering growth conditions to produce daughters of different sizes made it possible to show that daughters must reach a **threshold size** in order to pass Start: Smaller daughters must grow for a longer time before they can pass Start (Figure 3-13). In budding yeast and many other cells, a threshold size for Start is the principal means of coordinating the cell cycle engine and growth. But in some cells, like the fission yeast described in Chapter 4, the primary restraint on the cell cycle engine is a threshold size for entry into mitosis.

Although cells must reach a minimum size to go through some transitions in the cell cycle engine, no maximum size inevitably induces passage through Start or mitosis. Because the cell cycle can run faster than cells can grow, however, cells that become too big are quickly slimmed down. Budding yeast cells that are born bigger than the critical size required to pass through Start divide before they have doubled in mass. These short cell cycles continue until the cells are reduced to a size at which the mechanism that monitors cell size once more restrains the cell cycle engine.

Some cells, such as oocytes and large nerve cells, achieve enormous size by suppressing cell division while growth continues. The rapid early embryonic cell cycles are possible because the egg and early cleavage stage cells are much larger than the threshold sizes that regulate the cell cycle in somatic cells. Within a generation in an animal's life, the growth without division during oogenesis and the division without growth during embryogenesis exactly cancel each other out (Figure 3-14). As a result, the average size of a given cell type remains constant from generation to generation.

Start coordinates the cell cycle engine with nutrient availability

In the evolutionary struggle to perpetuate their genomes, organisms reproduce as fast as they can. Since in unicellular organisms, like yeasts, cell division and reproduction are one and the same thing, evolution has selected

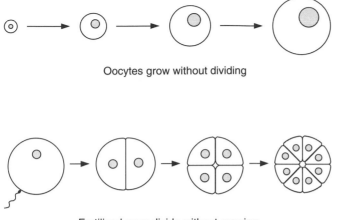

Oocytes grow without dividing

Fertilized eggs divide without growing

FIGURE 3-14 Cell size varies during development. Normal regulation of cell size is suspended during reproduction. Frog oocytes are born the same size as other cells but grow to be 1 mm in diameter without dividing. Once the oocytes have matured and been fertilized, they divide without growing, thus reducing the cell size.

cells that divide rapidly. In such organisms, starvation is one of the most common curbs on the cell cycle. Two mechanisms keep sudden removal of nutrients from leading to a rapid depletion of ATP that would stall the cell cycle engine. First, cells maintain internal stores of nutrients that are large enough to support DNA replication, mitosis, and cell division. Second, they respond to depletion of these stores by arresting the cell cycle at specific checkpoints. In some organisms, like budding yeast, mild starvation blocks passage through Start but has no effect on entry into mitosis. In others, like fission yeast, passage through Start and entry into mitosis are equally sensitive to starvation.

Using two checkpoints to respond to starvation probably enables cells to escape the lethal effects of errors in DNA replication and chromosome segregation. Since arrest during S phase leads to DNA damage, cells use Start as a checkpoint to make sure enough food is available to complete DNA replication. If nutrient supplies are low the cell cycle arrests to avoid stalling during this crucial process, thus minimizing the chance of genetic damage. In a similar manner, cells use the entry into mitosis as a checkpoint to prevent cells from stopping in mitosis and missegregating their chromosomes.

In warm-blooded, multicellular organisms, cells are bathed in plentiful nutrients and are kept at constant and optimal temperature. Although the absence of nutrients does arrest the cell cycle, their presence is not enough to induce cell division. Instead, most physiological control of cell growth and division is mediated by special polypeptide **growth factors** that regulate passage through the same cell cycle checkpoints that respond to simple nutrients in unicellular organisms. Growth factors dictate the carefully regulated patterns of cell division that produce and maintain the complex organs of multicellular organisms. Breakdowns in these elaborate regulatory mechanisms can lead to the uncontrolled cell proliferation that characterizes malignant cancer.

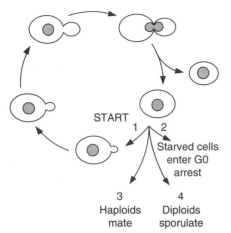

FIGURE 3-15 Alternative cell fates in budding yeast. After mitosis ends, budding yeast cells can do one of the following: (1) pass through Start, replicate their DNA, and undergo mitosis; (2) enter the resting G0 phase in response to starvation; (3) mate if haploid; (4) undergo meiosis and sporulate if diploid. Once cells have passed through Start, however, they are committed to the mitotic cell cycle and must exit mitosis before these other options are again possible.

Starvation can induce cells to enter a resting phase

The absence of nutrients or growth factors causes cells to enter specialized resting states. Yeast cells starved of nutrients or mammalian cells deprived of growth factors arrest early in G1 in a state called **G0** (Figure 3-15). The arrested cells have decreased rates of protein synthesis and often have increased resistance to temperature extremes and other environmental stresses. Entry into this specialized resting state can be thought of as a form of cell differentiation designed to ensure survival under adverse conditions.

Some budding yeast *cdc*^ts^ mutants have defects in the mechanisms that cells use to monitor nutrient availability. The *cdc25*^ts^ and *cdc35*^ts^ mutants block passage through Start, but unlike *cdc28*^ts^, these mutants also prevent cells from growing at the restrictive temperature and induce them to enter G0. The phenotypes of *cdc25*^ts^ and *cdc35*^ts^ are consequences of the fact that these mutations deprive cells of the ability to detect nutrients in their environment (see pages 96 – 100).

Until recently it was thought that only cells between mitosis and Start could enter specialized resting states. Recent evidence suggests, however, that just as treating yeast cells with mating factors at any point in the cell cycle activates a signaling pathway, so starvation at any point in the cell cycle induces a response that includes the cessation of growth, decreased protein synthesis, and increased stress resistance. But just like the signaling pathway activated by mating factors, starvation can arrest the cell cycle only at certain checkpoints.

Starving cells or depriving them of growth factors arrests different cells at different checkpoints. Budding yeast and mammalian tissue culture cells are arrested in G0. In other cells, such as immature oocytes, the checkpoint is entry into mitosis, and cells can be arrested in a specialized resting state with a G2 DNA content. Some somatic cells can also show this prolonged G2 arrest, confirming the idea that similar signals affect different checkpoints in different cells.

CONCLUSION

In this chapter we have seen that budding yeast uses Start as a checkpoint to coordinate the cell cycle with cell size and nutrient availability and to regulate entry into alternative cell cycles. Is Start a universal transition in somatic cell cycles? We cannot yet answer this question conclusively. Other unicellular eukaryotes have homologs of Cdc28 that induce passage through Start. In most vertebrate cells the restriction point fulfills the same role in regulating the cell cycle that Start plays in unicellular cells, but our ignorance about the molecules that control the restriction point makes it difficult to tell how closely related to Start this transition is.

SELECTED READINGS

Review

Hartwell, L. H. Twenty-five years of cell cycle genetics. *Genetics* **129**, 975–980 (1991). *An elegant historical review of the progress made by genetically analyzing the cell cycle.*

Original Articles

Hartwell, L. H., Culotti, J., & Reid, B. Genetic control of cell division in yeast, I. Detection of mutants. *Proc. Natl. Acad. Sci. USA* **66**, 352–359 (1970). *Isolation of the first cell cycle mutants.*

Hartwell, L. H., Culotti, J., Pringle, J. R., & Reid, B. J. Genetic control of cell division cycle in yeast. *Science* **183**, 46–51 (1974). *A paper that reviews the evidence that led to the idea of Start.*

Hartwell, L. H., & Unger, M. W. Unequal division in *Saccharomyces cerevisiae* and its implications for the control of cell division. *J. Cell. Biol.* **75**, 422–435 (1977). *Daughter cells must grow until they are large enough to pass Start.*

ENZYMES THAT CONTROL MITOSIS

*B*Y 1 9 8 4 research on yeast and frogs had generated two very different views of the cell cycle. Experiments on frog eggs suggested that the cell cycle was driven by a free-running cell cycle engine that produced regular oscillations in the activity of MPF, and the discovery of cyclin in sea urchin eggs suggested a biochemical model for the engine. In contrast, genetic analysis of the cell cycle in budding yeast had produced an elaborate logical map of the cell cycle with a complex web of interdependency among the different events within the cell cycle. How could the apparent discrepancy between the organization of the embryonic and somatic cell cycles be reconciled, and how could biochemical models for the cell cycle be tested? The answer was provided by a key advance that unified genetics and biochemistry: the discovery that MPF is composed of two subunits, cyclin B and the product of the *cdc2$^+$* gene, the master regulator of mitosis in the fission yeast *Schizosaccharomyces pombe*. This chapter retraces the steps that led to this discovery, beginning with a description of the life-style and genetics of fission yeast.

CONTROL OF MITOSIS IN FISSION YEAST

Fission yeast first attracted human attention as the agent that produced the African beer called *pombe*. Although fission yeast and budding yeast are both used in brewing, they are very different from each other. Comparing the nucleic acid sequences of the ribosomal RNAs of the two yeasts suggests that their most recent common ancestor lived more than a billion years ago. This is about the time that the lineage that would eventually lead to mammals split off from those leading to the two yeasts, suggesting that the two yeasts are no more related to each other than they are to us.

Fission yeast cells grow by elongation and divide by septation

As its name suggests, fission yeast grows and divides fundamentally differently from budding yeast. Fission yeast are rod-shaped cells that grow by elongation and divide by laying down a cell wall across the middle of the rod (Figures 4-1 and 4-2). During mitosis, the three fission yeast chromosomes condense to the extent that they can just be distinguished by light microscopy. In both yeasts the nuclear envelope remains intact throughout mitosis.

Both budding yeast and fission yeast can mate and sporulate, but they have different life-styles. In the wild, the haploid spores of budding yeast germinate and immediately mate with each other to form diploids that reproduce mitotically until starvation induces meiosis and sporulation (see Figure 3-9). By contrast, fission yeast spores germinate and divide as haploids. These cells do not mate with each other unless starved, and the resulting diploid cells sporulate as soon as they are formed (Figure 4-3). In the laboratory it is easy to make budding yeast to reproduce as haploids, but it is more difficult to make fission yeast reproduce as diploids.

Like budding yeast, fission yeast grows rapidly on simple media and is well suited for both classical and molecular genetic analysis. Because they do not form a bud, there is no simple marker for Start in fission yeast. Nevertheless, two properties of this organism make it especially suitable for analyzing

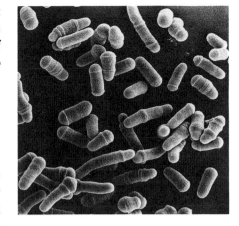

FIGURE 4-1 Fission yeast. This scanning electron micrograph shows a population of fission yeast cells. Note the uniform cell diameter, which makes it possible to estimate the mass of cells simply by measuring their length. Micrograph courtesy of N. Hajibagheri.

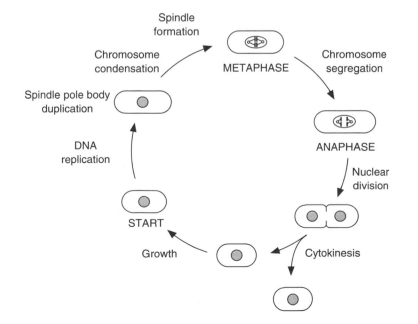

FIGURE 4-2 Fission yeast cell cycle. The cell cycle of fission yeast differs from that of the budding yeast in two ways: The fission yeast cycle has a clear G2 phase between the end of DNA replication and the onset of spindle assembly, and the budding yeast cycle does not; unlike budding yeast, in which budding indicates that cells have passed Start, fission yeast lacks a morphological marker for Start.

FIGURE 4-3 Fission yeast life cycle. Unlike budding yeast, which can divide as either a haploid or a diploid, fission yeast can divide only as a haploid. The haploid exists as one of two mating types, *h⁺* and *h⁻*. Cells of opposite mating types can mate with each other other when starved of nitrogen, and immediately after mating they enter a meiotic cell cycle and produce four haploid spores.

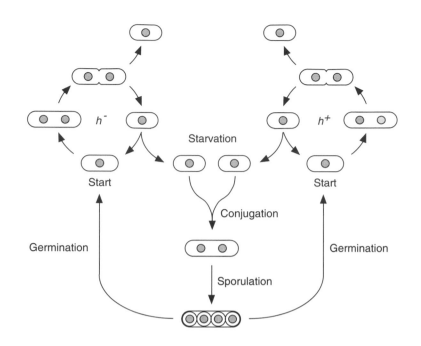

the relationship between cell size and the cell cycle. First, unlike budding yeast, cells divide symmetrically, so the size of cells at the same position in the cell cycle is very uniform. Second, fission yeast grows only in length and not in diameter. Measuring the length of individual cells thus provides an easy and precise way of following cell growth; such measurements permit detailed analysis of how cell size regulates progress through the cell cycle.

Fission yeast must reach a critical size to enter mitosis

Chapter 3 described how the requirement that cells reach a threshold size before passing Start coordinates the cell cycle engine of budding yeast with cell growth. Analysis of a fission yeast mutant that produced small cells revealed a size threshold that regulated entry into mitosis. The *wee1ᵗˢ* mutation (*wee* is Scottish for "small") was isolated as a recessive, temperature-sensitive mutation. At the permissive temperature of 25°C, the average cell length of *wee1ᵗˢ* cells was close to that of the wild type, but at 35°C it was only half that of wild-type cells. Since most recessive mutations inactivate proteins, the phenotype of *wee1ᵗˢ* cells suggests that the product of the *wee1⁺* gene restrains the cell cycle engine until cells reach a suitable size for division.

Although *wee1ᵗˢ* cells are smaller when grown at 35°C than they are when grown at 25°C, the duration of the cell cycle at both 25°C and 35°C is identical to that of a wild-type strain grown under the same conditions. When a culture of *wee1ᵗˢ* cells is shifted from 25°C to 35°C, the length of the cell cycle that includes the temperature shift is shorter than usual (Figure 4-4). This shortened transitional cell cycle shows that under certain conditions, a cycle can occur in less time than it takes cells to double their mass. This short

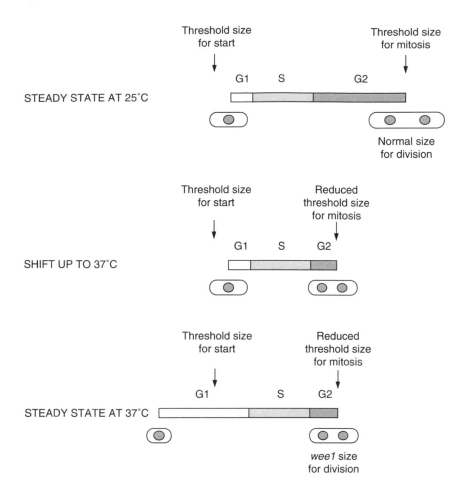

STEADY STATE AT 25°C

Threshold size for start

Threshold size for mitosis

G1   S   G2

Normal size for division

SHIFT UP TO 37°C

Threshold size for start

Reduced threshold size for mitosis

G1   S   G2

STEADY STATE AT 37°C

Threshold size for start

Reduced threshold size for mitosis

G1   S   G2

wee1 size for division

FIGURE 4-4   Minimum size for entry into mitosis. Fission yeast that carry the *wee1ts* mutation grow normally at the permissive temperature. When the cells are shifted to the restrictive temperature, however, the minimum size required to pass the mitotic entry checkpoint is greatly reduced. The length of the cell cycle in which the shift to the restrictive temperature occurred is thus dramatically reduced, since the cells no longer have to double in size in order to divide. However, subsequent cycles at the restrictive temperature are of normal duration, since the cells, now smaller at birth, must again double in size before they can divide. Reference: Nurse, P. *Nature* **256**, 547–551 (1975).

cycle also implies that the normal function of the *wee1+* gene is to delay a transition in the cell cycle engine until cells have reached a critical size.

How can we tell which transition is inhibited by the Wee1 protein? This question was answered by an experiment in which the amount of DNA and the number of cells in a *wee1ts* culture were measured continuously during the transitional cell cycle. If the Wee1 protein inhibited Start, inactivating it by shifting the *wee1ts* cells to 35°C would reduce the threshold size for passing Start and lead to a sudden burst of DNA synthesis. If, instead, Wee1 inhibited mitosis, inactivating the protein would reduce the threshold size for mitosis and lead to a sudden burst of increased cell division. Careful analysis of the transitional cycle showed such a burst of increased cell division, without an accompanying burst of DNA synthesis, demonstrating that Wee1 inhibits entry into mitosis until cells reach a threshold size.

The Wee1 protein is part of a checkpoint that keeps the cell cycle engine from running faster than cells can grow and acts as a homeostatic mechanism to maintain constant cell size. Cells born smaller than normal have long cell cycles because they must more than double their mass to reach the threshold size. Cells born larger than normal divide early because they do not have to double their mass to reach the threshold size. Thus, the size checkpoint ensures that even if cells are born smaller or larger than normal, they will produce normal-size progeny.

Fission yeast must also reach a critical size to pass Start

A detailed analysis of *wee1^ts* mutants revealed not one but two points in the fission yeast cell cycle at which the cell cycle engine is coordinated with cell growth. At 25°C Start occurs very early in the cell cycle of both *wee1^ts* and wild-type cells, so that cells begin DNA replication as soon as they finish mitosis. At 35°C, however, *wee1^ts* mutants do not begin DNA replication until they are more than halfway through the cell cycle. The existence of two threshold sizes, one to pass Start and another to enter mitosis, explains this finding. In wild-type cells, the threshold size for Start is about the size at which cells are born, so they are already big enough to pass Start and begin DNA replication when mitosis finishes. At 35°C the *wee1^ts* mutant greatly reduces the threshold size for entry into mitosis but does not change the size required to pass Start. As a result the *wee1^ts* cells are born considerably smaller than the critical size required to pass Start and must grow before they can begin DNA replication (see Figure 4-4).

The conclusion that cells must reach critical sizes both to pass Start and to enter mitosis also applies to budding yeast and probably to all growing cells. In a given cell type, however, usually only one of the control points is visible. In wild-type fission yeast, cells are born large enough to pass Start, so this size control is invisible. Conversely, by the time wild-type budding yeast cells are big enough to pass Start, they are also big enough to enter mitosis, so in this case the size control for entry into mitosis is invisible.

Mutations that reduce the threshold size for the major checkpoint expose the minor one. In fission yeast, *wee1^ts* mutants reveal the existence of a threshold size for Start. In budding yeast, mutants that reduce the critical size for passage through Start reveal the existence of a threshold size for mitosis; the interval between Start and mitosis increases because cells must now grow to reach the critical size required for entry into mitosis.

The fission yeast *cdc2^ts* mutation identifies a critical inducer of mitosis

Just as the genetic analysis of *cdc^ts* mutants in budding yeast identified *CDC28^+* as a key inducer of Start, so the analysis of *cdc^ts* mutants in fission yeast identified a central regulator of mitosis. The fission yeast *cdc^ts* mutants were isolated as temperature-sensitive mutations that failed to divide but continued to grow, forming long rods, at the restrictive temperature. Analyzing the DNA content of mutant cells and their ability to synthesize DNA at the restrictive temperature allowed the *cdc*^{ts} mutants to be classified according to the point at which they arrested the cell cycle: G1, S phase, or G2.

The *cdc*^{ts} mutations that prevented entry into mitosis were particularly informative. The phenotypes of different mutations of *cdc2*$^+$ suggested that this gene encodes the key protein that induces mitosis. Most mutations in *cdc2*$^+$ were recessive, temperature-sensitive mutations that blocked entry into mitosis at the restrictive temperature. Revealingly, a nonconditional, dominant mutation, called *cdc2-3w*^{D}, had exactly the opposite effect. Instead of preventing cells from entering mitosis, it allowed them to go through mitosis and divide at a smaller size than wild-type cells. Dominant mutations tend to increase the activity of proteins either by increasing their level in the cell or by freeing them from negative regulation by other proteins. The phenotypes of the two classes of mutant showed that removing Cdc2 activity kept cells from entering mitosis, whereas increasing it accelerated entry into mitosis.

Unfortunately, the numbering of *cdc*^{ts} mutants in fission yeast bears no relationship to that in budding yeast, leading to great confusion. For example, as we shall see, the *CDC28*$^+$ gene of budding yeast is the same gene as the *cdc2*$^+$ gene of fission yeast, whereas the *CDC25*$^+$ gene of budding yeast and the *cdc25*$^+$ gene of fission yeast encode unrelated genes that perform completely different functions. The Appendix lists the *cdc* mutations of budding and fission yeast and what is known about each of them.

The mitosis-inducing activity of Cdc2 is inhibited by Wee1 and stimulated by Cdc25

Studying the interactions of the different mutants that affected entry into mitosis suggested that *cdc2*$^+$ encoded a key component of the cell cycle engine, whose activity was regulated by other proteins. No other mutation could compensate for the loss of Cdc2 activity, identifying it as an indispensable component of the cell cycle engine. By contrast, the *cdc25*^{ts} mutation, which arrests cells at exactly the same point in the cell cycle as *cdc2*^{ts}, can be rescued by the *wee1*^{ts} mutation. The growth and division of *cdc25*^{ts} *wee1*^{ts} double mutants are normal at the restrictive temperature, suggesting that the function of Cdc25 is to overcome the ability of Wee1 to inhibit entry into mitosis (Figure 4-5).

An appealing explanation of the interactions of the various mutants is that Cdc25 activates the mitosis-inducing activity of Cdc2 and Wee1 inhibits it (Figure 4-6). If Cdc25 activated Cdc2, it would explain why *cdc25*^{ts} mutations cannot enter mitosis at the restrictive temperature. The ability of Wee1 to inhibit Cdc2 activity would explain why *wee1*^{ts} mutants, like *cdc2-3w*^{D}, enter mitosis prematurely. Finally, the antagonistic actions of Wee1 and Cdc25 on Cdc2 would explain why *wee1*^{ts} mutations allow *cdc25*^{ts} mutants to enter mitosis. Subsequent biochemical experiments spectacularly verified this genetically derived model for the control of mitosis (pages 57–59).

The combination of the *cdc2-3w*^{D} and *wee1*^{ts} mutations showed that the

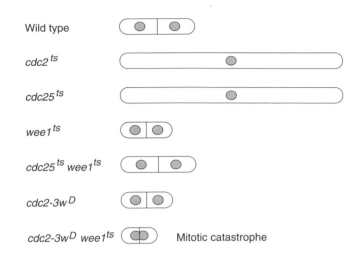

FIGURE 4-5 Fission yeast mutants that affect cell size. The diagram depicts the size of wild-type and mutant cells at the end of mitosis. The *cdc2ᵗˢ* and *cdc25ᵗˢ* mutations block entry into mitosis at the restrictive temperature; the *wee1ᵗˢ* and dominant *cdc2-3wᴰ* mutations allow cells to enter mitosis at a smaller size than mitosis in wild-type cells. At the restrictive temperature, the *cdc2-3wᴰ wee1ᵗˢ* mutant divides at a size that is small enough to cause cell death. Reference: Russell, P., & Nurse, P. *Cell* **49**, 559–567 (1987).

regulation of the cell cycle engine can become so deranged that it kills cells. Both of these mutations accelerate entry into mitosis. At the permissive temperature for the *wee1ᵗˢ* mutation, *cdc2-3wᴰ wee1ᵗˢ* double mutant cells are smaller than wild-type cells but divide to produce viable cells. But when the cells are shifted to the restrictive temperature they suffer **mitotic catastrophe:** They divide at an extremely small size and die, providing a clear demonstration of the need to coordinate the cell cycle engine with cell growth and the completion of downstream events.

Cloning the *cdc2⁺* gene showed that it encodes a protein kinase

Although genetic analysis identified the fission yeast genes that induce mitosis, it could not disclose the functions of the proteins they encoded. Fortunately, advances in genetic engineering made it possible to isolate these genes, determine their nucleotide sequence, and thus predict the amino acid sequence of the proteins that they encode. This approach can predict the function of the new protein if its sequence resembles that of proteins whose biochemical functions are already known.

Cloning the wild-type fission yeast *cdc2⁺* gene yielded a valuable clue about the cell cycle engine. The gene was cloned by isolating a DNA fragment that would complement a *cdc2ᵗˢ* mutation. This general approach works well for both fission and budding yeast and uses libraries of DNA plasmids that can replicate in both yeast and bacteria. Each member of the library contains a different fragment of yeast DNA. To clone the fission yeast *cdc2⁺* gene, a library of plasmids containing fission yeast DNA fragments was introduced into *cdc2ᵗˢ* cells, which were then incubated at 35°C. Each recipient cell received a different plasmid from the library. The vast majority of cells received a plasmid that did not contain the *cdc2⁺* gene and failed to grow into colonies at 35°C, but a tiny fraction of the cells received a plasmid that did

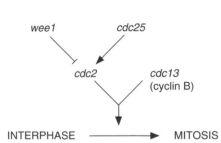

FIGURE 4-6 Control of mitosis in fission yeast. The logical interactions of the different genes that regulate entry into mitosis in fission yeast. Both *cdc2* and *cdc13* are absolutely required for entry into mitosis. The ability of *cdc2* to induce entry into mitosis is stimulated by *cdc25* and inhibited by *wee1*.

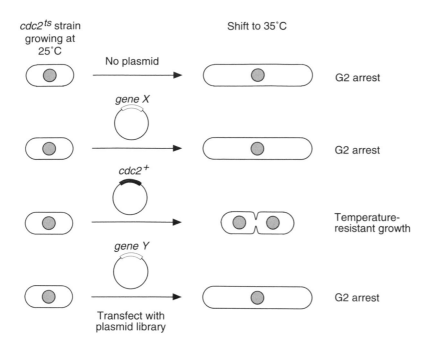

FIGURE 4-7 Cloning genes by complementing mutations. A library of plasmids carrying fragments of fission yeast DNA is transfected into a *cdc2ts* mutant of fission yeast. Each plasmid in the library contains one fragment of fission yeast DNA, and each fission yeast cell receives only one plasmid from the library. Most cells get a plasmid carrying a gene unrelated to *cdc2+* and cannot grow at the restrictive temperature. A very small fraction of cells get a plasmid carrying the wild-type *cdc2+* gene. The plasmid can be recovered from these cells and used to analyze the gene.

contain the *cdc2+* gene and did give rise to colonies (Figure 4-7). The plasmid that these cells contained was recovered, amplified in bacteria, and used to determine the nucleotide sequence of the *cdc2+* gene.

The amino acid sequence of the Cdc2 protein that was deduced from the gene's nucleotide sequence showed homology to protein kinases (Figure 4-8). This clue was exciting, because the pattern of protein phosphorylation was

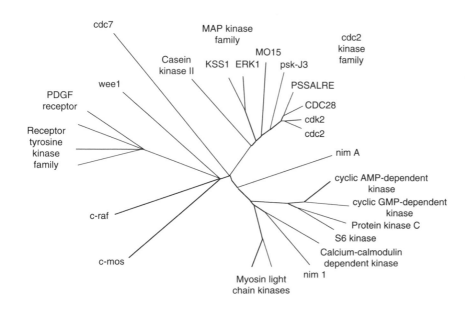

FIGURE 4-8 Evolutionary relationships among protein kinases. This diagram places many of the protein kinases involved in the cell cycle on an evolutionary tree. At each branch point, the length of the branches indicates the relatedness of the protein kinases in that family; the longer the branch the more distantly related the members of the family. Thus Cdc2 and Cdc28 are closely related to each other, while Cdc7 (a kinase required for the initiation of DNA synthesis in budding yeast) is only distantly related to Wee1. The lengths of the lines connecting two different families of kinases indicate their relatedness to each other; the cyclic AMP- and cyclic GMP-dependent kinases are closely related to the protein kinase C family, but both of these families are only distantly related to the tyrosine kinases.

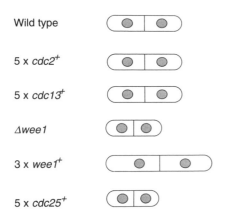

FIGURE 4-9   Cell size control in fission yeast. The sizes of cells at mitosis with different doses of the *wee1+* and *cdc25+* genes are shown. Increasing the dosage of *wee1+* increases the size that cells must reach before entering mitosis, suggesting that Wee1 inhibits entry into mitosis. Increasing the dosage of *cdc25+* decreases the minimum size required for mitosis, suggesting that Cdc25 activates mitosis. Increasing the dosage of *cdc2+* or *cdc13+* does not make cells smaller, showing that the level of their protein products does not normally determine when cells enter mitosis. Reference: Russell, P., & Nurse, P. *Cell* **49,** 559–567 (1987).

known to change when cells entered mitosis, and it had long been suspected that MPF was a protein kinase. Cloning the *wee1+* gene revealed that it too is homologous to protein kinases, and we now know that the Cdc25 protein is a protein phosphatase.

The amounts of the Wee1 and Cdc25 proteins regulate the entry into mitosis

Is it a change in the amount of Cdc2 protein or a change in its activity that induces mitosis? This question was addressed by introducing extra copies of the cloned *cdc2+* gene into cells to increase the amount of Cdc2 protein in them. The additional copies of the cloned *cdc2+* gene did increase the level of Cdc2 protein but had no effect on the cell cycle, showing that it is the activity rather than the amount of the Cdc2 protein that determines when mitosis occurs (Figure 4-9).

By contrast, overexpressing the *wee1+* and *cdc25+* genes confirmed the hypothesis that they control the speed of the cell cycle engine. Cells got bigger as the dosage of the *wee1+* gene was increased and smaller as the dosage of the *cdc25+* gene was increased. These experiments reveal that, unlike the situation with Cdc2, the amount of the Wee1 and Cdc25 proteins plays an important role in controlling the cell cycle. We shall see that the activity of these proteins is controlled by post-translational changes that affect their activity, as well as by mechanisms that regulate their abundance (pages 59–61).

A UNIVERSAL CELL CYCLE ENGINE

Although experiments on fission yeast had provided an elegant framework for thinking about the control of mitosis, it was unclear how far these conclusions could be extended to other organisms. This question was addressed by asking whether other eukaryotes contain homologs of the genes that control the cell cycle in fission yeast.

Cdc2/28 induces both Start and mitosis in yeasts

To determine if budding yeast has a homolog of the fission yeast *cdc2+* gene, a plasmid library containing fragments of budding yeast genomic DNA was introduced into a fission yeast strain carrying a *cdc2ts* mutation. Colonies that could grow at 35°C were selected, and the plasmids they contained were analyzed. Remarkably, the gene that complemented the fission yeast *cdc2ts* mutant was the budding yeast *CDC28+* gene, which had been defined as the critical gene required for budding yeast to pass Start (see Chapter 3). The sequence of the budding yeast *CDC28+* gene is 63% identical to that of the fission yeast *cdc2+* gene, showing that the two genes are the strongly conserved descendants of a single ancestral gene.

The finding that a budding yeast gene that induces passage through Start is homologous to a fission yeast gene that induces mitosis was initially hard to explain. Closer inspection revealed that the phenotypes of the *cdc28^ts* mutation in budding yeast and those of the *cdc2^ts* mutation in fission yeast correspond to the relative importance of Start and entry into mitosis for regulating the cell cycle engine in these two organisms. In budding yeast, Start is the key checkpoint for regulating the cell cycle in response to cell size, nutrient availability, and sex. Although Cdc28 activity was first thought to affect only Start, subsequent analysis showed that it was also required at mitosis. In fission yeast, entry into mitosis is the key checkpoint for cell-size control. Cdc2 activity was first thought to affect only mitosis, but more detailed analysis showed that it was also required to pass Start. Thus, despite the apparent complexity of the cell cycle, a single protein acts as a critical regulator of the two key transitions in the cell cycle of both budding and fission yeast. We refer to the protein product of this gene as Cdc2/28 unless specifically discussing budding or fission yeast. Because of its molecular weight, Cdc2/28 is often called p34^cdc2 or simply p34.

The differences in their life cycles, coupled with the need for cells to avoid DNA damage, may explain why budding yeast spends much longer in G1 than does fission yeast. In the wild, budding yeast grows as a diploid and fission yeast grows as a haploid. Ultraviolet light, X rays, and mutagenic chemicals can break both strands of a DNA molecule. In both yeasts, these breaks can be repaired only by using an intact chromosome as a template to restore a broken one.

In a haploid cell, the template used for repairing double-stranded breaks is a sister chromatid. Because sister chromatids do not exist until DNA replicates, haploid cells cannot repair DNA damage that occurs before DNA replication. Thus, organisms that grow as haploids, like fission yeast, are under strong selective pressure to replicate their DNA as early in the cell cycle as possible to minimize their vulnerability to DNA damage. Diploid cells, however, can repair DNA damage in G1 because there are two copies of each chromosome, so any damaged chromosome has an intact partner that it can use as a template for repair. Therefore, there is no special selective pressure for diploid organisms, like budding yeast, to have a short G1.

All eukaryotes have a functional homolog of *cdc2*

Cloning a human *cdc2^+* gene that complements a fission yeast *cdc2^ts* mutation provided the most impressive demonstration of the evolutionary and functional conservation of the cell cycle engine. This exciting discovery suggested that the fundamental features of cell cycle control are the same in all eukaryotes. In addition, it showed that a key component of the cell cycle engine has been so highly conserved in evolution that the human version can be plugged into the yeast cell cycle engine without disturbing its operation. We now know that all eukaryotes contain functional homologs of *cdc2^+*, in which about 65% of the amino acid sequence is identical to that of the fission yeast *cdc2^+* gene and the budding yeast *CDC28^+* gene.

MPF is a protein kinase composed of Cdc2/28 and cyclin B

By 1988 a wide variety of studies on the induction of mitosis were poised for unification. Genetics had shown that Cdc2/28 is present in all eukaryotes, and physiology and biochemistry had shown that MPF is present in all mitotic and meiotic cells. The rise and fall of cyclin during embryonic cell cycles strongly suggested that it plays a key role in inducing mitosis. Were Cdc2/28, cyclin, and MPF all part of a single entity that regulated mitosis or were they branches of separate pathways that converged to induce mitosis?

The golden spike that united the different approaches to mitosis was the biochemical purification of MPF. After years of effort, methods were developed that could produce a few micrograms of MPF from 1 liter of frog eggs. As expected from earlier work, the purified MPF was a protein kinase, but the key question was the identity of its two subunits. Their molecular weights led to speculation that these two proteins were Cdc2/28 and cyclin B. This prediction was rapidly and gratifyingly confirmed when antibodies raised against a highly conserved amino acid sequence in Cdc2/28 recognized the 34 kd (kilodalton) subunit of MPF, and antibodies to frog cyclin B recognized the 46 kd subunit.

Physical and biochemical studies of purified MPF have given us more information about this universal inducer of mitosis. Pure MPF is a heterodimer containing one cyclin and one Cdc2/28 subunit, although in crude extracts MPF activity is associated with larger protein complexes. One of the best substrates for MPF is histone H1, and its activity is commonly measured as the H1 kinase activity in crude extracts from cells. In fact, MPF is very similar to a protein kinase called growth-associated histone H1 kinase, which, like MPF, shows a large increase in activity during mitosis and had resisted purification for many years. Two forms of cyclin B, called B1 and B2, can associate with Cdc2/28 to form active kinase complexes. Some subtle differences between the substrate specificities of these different kinases have been demonstrated, but their biological significance remains unclear.

CYCLIN AND THE CELL CYCLE

In the simplest model of the cell cycle engine (see Figure 2-13), newly synthesized cyclin B binds to Cdc2/28 and produces active MPF, and cyclin degradation at the end of mitosis inactivates MPF. This model predicts that cyclin would be the only protein that embryonic cells have to synthesize to induce mitosis and that cyclin degradation is required for the inactivation of MPF.

Cell cycle extracts show that cyclin synthesis drives early embryonic cells into mitosis

To test the predictions of the simple cell cycle model, it was necessary to specifically control the synthesis and degradation of cyclin. The only way to do this in embryonic cell cycles was to develop extracts that perform the cell

FIGURE 4-10 Cell cycle extracts. Techniques have been developed to break frog eggs gently and recover their undiluted cytoplasm. Frog eggs are electrically activated to restart the cell cycle, packed to remove excess buffer, and then crushed by spinning them at 10,000 *g*. The centrifugation separates the contents of the eggs into three layers: lipid droplets on top, yolk particles on the bottom, and cytoplasm in the middle. The cytoplasmic layer is recovered, allowing the cell cycle to be recreated in vitro.

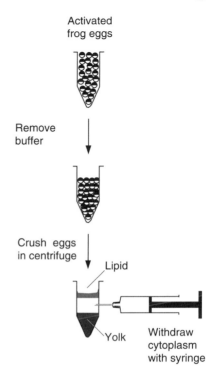

cycle in a test tube, where it is more accessible to experimental manipulation. The cell cycle extracts consist of pure cytoplasm prepared from frog eggs. Unfertilized eggs are activated with an electric shock that, like fertilization, increases the intracellular calcium concentration and releases the eggs from their arrest in metaphase of meiosis II, initiating the mitotic cell cycles. The eggs are concentrated and then gently broken by centrifuging them at high speed, which also separates their contents into three layers: a pellet of yolk particles, a floating layer of lipid, and sandwiched between them, crude and undiluted cytoplasm (Figure 4-10). When sperm nuclei are added to the cytoplasm, they swell, acquire a nuclear envelope, and replicate their DNA (Figure 4-11). Because the cytoplasm synthesizes proteins, cyclin accumulates during interphase leading to MPF activation, chromosome condensation, and nuclear envelope breakdown, just as it does in an intact egg. Within 10 minutes of nuclear envelope breakdown, cyclin is suddenly degraded, MPF activity falls, the chromosomes decondense, and interphase nuclei reform. These extracts can perform multiple cell cycles, with a cycle length only twice that of intact frog eggs.

   Inhibition of protein synthesis stops the cell cycle in these extracts, just as it does in fertilized eggs. The added nuclei swell and replicate their DNA, but they never enter mitosis. The pattern of protein synthesis in the cell cycle extracts is very complex, and any number of the newly made proteins could be required to induce mitosis. To see if cyclin synthesis alone could induce mitosis, it was necessary to develop a method to inhibit the synthesis of all proteins except cyclin. This aim was achieved in three steps. First, the extracts were treated with low doses of ribonuclease, which destroyed the mRNA in the extract without significantly damaging the rRNA and tRNA that carry out the actual catalytic steps of protein synthesis. Next, a powerful, specific inhibitor of ribonuclease was added, followed by pure sea urchin cyclin B mRNA that had been transcribed by bacterial RNA polymerase from a cloned cyclin gene.

FIGURE 4-11 Cell cycle in vitro. Cytoplasmic extracts of frog eggs, prepared as shown in Figure 4-10, are mixed with sperm nuclei whose plasma membranes have been removed. On incubation, the sperm nuclei decondense into interphase nuclei and replicate their DNA. Later the extracts enter mitosis, and the nuclei break down and chromosomes condense, although classical spindles do not form. At the end of mitosis, MPF is inactivated, and the individual chromosomes decondense before fusing with each other to recreate interphase nuclei. These extracts perform from three to five cell cycles in vitro.

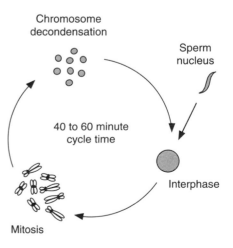

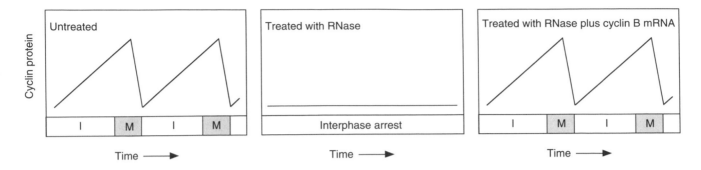

FIGURE 4-12 Cyclin synthesis drives the cell cycle. The behavior of three cell cycle extracts is shown. The left panel shows the control extract, which exhibits regular fluctuations in cyclin abundance coupled to the alternation of interphase (I) and mitosis (M). The right panel shows that destruction of all the mRNAs in the extract arrests the cell cycle in interphase. The bottom panel shows an extract treated sequentially with RNase to destroy the endogenous mRNA, an RNase inhibitor to block further RNase action, and purified cyclin B mRNA. Addition of the cyclin mRNA restores both cyclin synthesis and the cell cycle. Reference: Murray, A. W., & Kirschner, M. W. *Nature* **339**, 275–280 (1989).

Remarkably, the experiment worked. Ribonuclease treatment arrested the cell cycle of the extracts in interphase, and addition of the cyclin B mRNA restored the ability of the extracts to perform the cell cycle. During these cycles, cyclin accumulated in interphase and was destroyed at the end of mitosis, just as in living cells (Figure 4-12). The rate of the cyclin-driven cell cycles depended on the rate of cyclin synthesis; up to a point, the higher the concentration of cyclin mRNA, the faster the cell cycle. The ability of a sea urchin cyclin to drive the frog egg cell cycle showed once more how strongly evolution has conserved the fundamental components of the cell cycle engine.

Although cyclin synthesis can induce mitosis, other proteins might be able to fulfill the same role. The ability of specific DNA oligonucleotides to block the synthesis of a single protein made it possible to show, however, that cyclin B is the only protein whose synthesis can induce mitosis. Oligonucleotides complementary to cyclin B mRNAs were added to a cell cycle extract made from frog eggs. The oligonucleotides annealed to the cyclin mRNA, and an enzyme called ribonuclease H (RNase H), which cleaves only RNA annealed to DNA, cut the RNA portion of the hybrid, specifically preventing the mRNA from being translated to produce cyclin protein (Figure 4-13). The destruction of cyclin B mRNA arrested the cell cycle in interphase, showing that cyclin B synthesis is necessary for entry into mitosis (Figure 4-14).

Cyclin degradation is required for inactivation of MPF and exit from mitosis

The simple model of the cell cycle proposes that accumulation of cyclin activates MPF, and degradation of cyclin inactivates MPF. Therefore, blocking cyclin degradation should prevent the inactivation of MPF and arrest cells in interphase. This prediction was verified using a mutant form of cyclin that could induce mitosis but could not be degraded (Figure 4-15). The mutant protein lacked the 90 amino-terminal amino acids of cyclin. When the mRNA for this truncated cyclin was added to an interphase extract, its translation

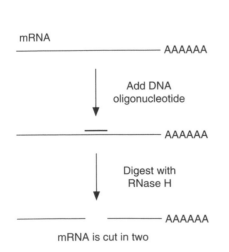

FIGURE 4-13 Specific destruction of an mRNA. A short (10 to 20 bases long) DNA oligonucleotide hybridizes to the mRNA that is complementary to it. Ribonuclease H (RNase H) specifically degrades the RNA portion of the RNA–DNA hybrid, cutting the message in half and blocking the synthesis of its protein product.

induced mitosis as usual, but the truncated cyclin was not degraded, and the extracts arrested in mitosis instead of returning to interphase.

Experiments with cell-free extracts thus confirmed the predictions of a model for the cell cycle engine in which cyclin synthesis drives cells into mitosis and cyclin degradation drives them into the next interphase. The ability to dissect the cell cycle engine in a test tube opened the way to biochemically fractionating the cell cycle engine and ultimately rebuilding it from purified components. Also, the ability of these extracts to reproduce the downstream events of the cell cycle makes biochemical dissection of these events possible. In addition to replicating DNA, cell cycle extracts can assemble mitotic spindles, separate sister chromatids, and perform anaphase chromosome movement (see pages 83 – 85).

Post-translational events are required for entry into mitosis

The rate of cyclin synthesis is not the only factor that controls the speed of the cell cycle engine in frog eggs. Although reducing the normal rate of cyclin synthesis can slow the cell cycle down, increasing it hardly speeds the engine up at all, suggesting that there must be other steps that limit the rate of MPF activation. These steps were also revealed by adding protein synthesis inhibitors to fertilized sea urchin eggs at different times after fertilization. Adding the inhibitors early in interphase blocked the next mitosis, but adding them during the second half of interphase did not (Figure 4-16). Therefore, during the first half of interphase cells made enough cyclin to induce mitosis, showing that the activation of MPF requires post-translational steps that occur well

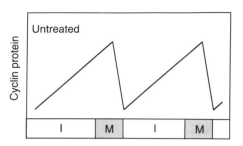
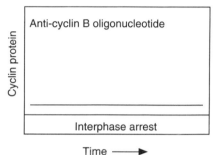

FIGURE 4-14 Cyclin synthesis is essential for the cell cycle. Frog cell cycle extracts were treated with antisense oligonucleotides that induce the destruction of cyclin-B1 and cyclin-B2 mRNAs. This treatment inhibits the synthesis of any cyclin B and arrests the extract. This experiment leads to the conclusion that the synthesis of cyclin B is essential for the activation of MPF. Reference: Minshull, J., Blow, J. J., & Hunt, T. *Cell* **56**, 947 – 956 (1989).

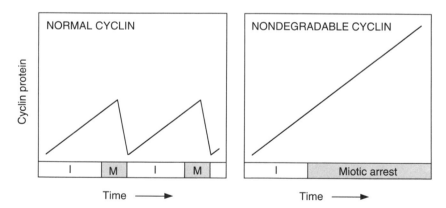

FIGURE 4-15 Cyclin destruction triggers the end of mitosis. Deleting the 90 N-terminal amino acids from cyclin B produces a mutant cyclin that can activate MPF but cannot be degraded. The left panel shows a control extract to which mRNA for wild-type cyclin B has been added. The right panel shows an extract to which mRNA for the truncated cyclin B has been added. Translating mRNA for the mutant cyclin in a cycling extract allows the extract to enter mitosis (M) but not to exit mitosis and reenter interphase(I), demonstrating that cyclin must be destroyed before MPF can be inactivated. Reference: Murray, A. W., Solomon, M. J., & Kirschner, M. W. *Nature* **339**, 280 – 286 (1989).

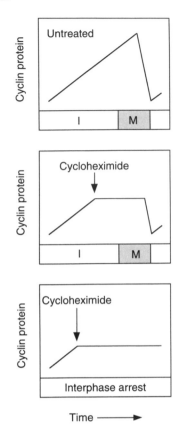

FIGURE 4-16 Post-translational steps in MPF activation. The effects of adding cycloheximide, an inhibitor of protein synthesis, at different times during interphase (I) of the first cell cycle of fertilized sea urchin eggs. The level of cyclin during the cell cycle is shown. Adding cycloheximide early in interphase (bottom panel) blocks entry into mitosis (M), but adding it in the second half of interphase (middle panel) does not. This experiment shows that protein synthesis is needed for entry into mitosis, but that steps that occur after cyclin has been made also limit the rate at which MPF is activated. Reference: Wagenaar, E. B. *Exp. Cell Res.* **144**, 393–403 (1983).

after cyclin has accumulated past the critical threshold. Interphase of the early embryonic cell cycle is thus divided into two periods: an early period in which cyclin accumulates to a critical threshold and a later period of post-translational events required to activate MPF. The rate of cyclin synthesis probably determines the length of the first period but not that of the second.

All eukaryotic cell cycles require cyclin synthesis, but its rate does not usually control the timing of mitosis. For example, in fission yeast, changing the dosages of the *cdc25⁺* and *wee1⁺* genes alters the timing of mitosis (see Figure 4-11), but changing the dosage of the *cdc13⁺* gene, which encodes cyclin B, does not. The amount of the Cdc25 protein in the fission yeast cell cycle fluctuates in a manner similar to that of cyclin: The concentration of the Cdc25 protein increases during interphase and decreases at mitosis. A similar result is seen in *Drosophila,* where the timing of certain mitotic divisions in the fertilized egg is determined by the timing of the synthesis of the *Drosophila* homolog of Cdc25. In early frog embryos, however, the amount of Cdc25 protein is constant throughout the cell cycle, although the activity of the protein varies. Thus, in different cells the behavior of the same protein may be regulated in different ways, and different proteins may act as key determinants of the timing of the cell cycle.

Oocyte maturation in clams and starfish underlines the importance of post-translational events for MPF activation and inactivation. Fertilization induces clam oocytes to mature: a hormonal signal induces the maturation of starfish oocytes. In both cases, the signal induces the conversion of preMPF into active MPF within 10 minutes, even in the complete absence of protein synthesis. Spindle assembly, the inactivation of MPF, and chromosome segregation also occur normally, showing that all these processes can be driven solely by modifications in the activities of preexisting proteins (see Chapter 5).

POST-TRANSLATIONAL REGULATION OF MITOSIS

This section describes experiments that identified the biochemical pathways that activate and inactivate MPF. The ultimate aim of these studies is to enumerate the components of the cell cycle engine, understand their activities, and reconstitute a working engine from purified components. A more modest goal is to ask whether our current understanding of the cell cycle engine can be embodied in mathematical models that produce a cell cycle on paper resembling the one we can monitor inside cells.

Protein phosphorylation regulates the activity of proteins in many ways

Protein phosphorylation is the most common post-translational modification that regulates processes inside cells and plays a key role in regulating the cell cycle engine. Protein kinases add phosphates to proteins by transferring phosphate groups from ATP to hydroxyl groups on amino acid side chains; protein phosphatases remove the phosphate groups.

The protein kinases and phosphatases that govern the eukaryotic cell cycle are divided into three classes based on the amino acids that they phosphorylate and dephosphorylate. One class recognizes serine or threonine residues, another recognizes tyrosine residues, and a small group recognizes serine, threonine, and tyrosine residues. Estimates suggest that eukaryotic cells have about 500 different protein kinases but many fewer different phosphatases. Some protein kinases, like MPF, phosphorylate many different substrates; others, like Wee1, act on only a single known substrate.

Phosphorylation of a given amino acid in a protein can have a variety of effects: activating or inactivating a protein's enzymatic activity, or increasing or decreasing a protein's affinity for binding to other proteins. Phosphorylation by one kinase can alter the ability of other kinases or phosphatases to catalyze reactions at nearby amino acids, making regulation of target proteins dependent on inputs from more than one signaling pathway.

Proteins can have multiple sites of phosphorylation that act as substrates for different protein kinases. Phosphorylating different sites in a protein can have dramatically different consequences. For example, as we shall see, phosphorylation on one site in Cdc2/28 inhibits the kinase activity of MPF, while phosphorylation on another site is absolutely required for kinase activity.

Protein kinases can interact with their substrates to produce complex control circuits that regulate intracellular processes. In some cases, regulation occurs via protein kinase cascades, in which protein kinases phosphorylate other protein kinases, thereby regulating their activity. Such cascades can contain several protein kinases that amplify a small initial signal into a large biochemical or structural change and play an important part in the regulation of the cell cycle by extracellular signals (see Chapter 6).

Technical problems make it more difficult to study protein phosphatases than protein kinases, and we still have much to learn about their variety and regulation. Serine/threonine phosphatases are divided into four families on the basis of their substrate specificity and their regulation, but each of these families probably contains many members with diverse physiological roles. Protein phosphatases are also subject to complex regulation involving phosphorylation and dephosphorylation and combination with specific inhibitors and regulatory subunits.

Phosphorylation of different sites on Cdc2/28 can either stimulate or inhibit MPF activity

Like many other protein kinases, the activity of MPF is regulated by phosphorylation. The Cdc2/28 subunit undergoes changes in phosphorylation on tyrosine and threonine residues during the cell cycle. The key phosphorylated residues are tyrosine 15 and threonine 161 (we use the numbering of the amino acids in the human Cdc2/28 protein). Phosphorylation of threonine 161 is required for kinase activity; phosphorylation of tyrosine 15 inhibits the kinase activity of Cdc2/28. The inhibitory phosphorylation is dominant to the

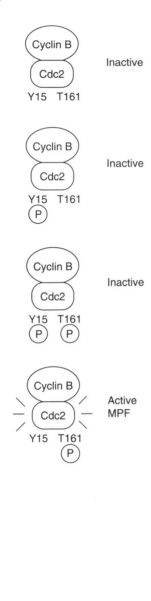

FIGURE 4-17 Control of MPF activity by Cdc2/28 phosphorylation. Cdc2/28 is phosphorylated (P) on tyrosine 15 (Y15) and threonine 161 (T161). Either phosphorylation can occur only on Cdc2/28 molecules that are bound to cyclin. Phosphorylation of tyrosine 15 inhibits the activity of the enzyme; phosphorylation of threonine 161, however, is necessary for MPF activity. When both tyrosine 15 and threonine 161 are phosphorylated, Cdc2/28 lacks kinase activity, showing the dominance of the inhibitory phosphorylation over the activating one.

activating one: Cdc2/28 phosphorylated on tyrosine 15 and threonine 161 lacks kinase activity (Figure 4-17).

How can we determine what effect phosphorylating a single amino acid on a particular protein, such as tyrosine 15 of Cdc2/28, has on a cell? A cell possesses many different protein kinases and phosphatases, and the majority of these have multiple protein substrates. This complexity makes it impossible to control the phosphorylation of a single amino acid in intact cells or crude extracts. Fortunately, advances in genetic engineering have made it possible to create mutant proteins that cannot be phosphorylated on particular amino acids and introduce them into cells. This aim is achieved in three steps. First, the phosphorylated site is identified on the protein. Then a mutant gene that can replace the phosphorylated amino acid with an amino acid that cannot be phosphorylated is engineered. (Alanine replaces serine or threonine residues, and phenylalanine replaces tyrosine residues.) Finally, the mutant gene is introduced into cells to determine the effect of blocking phosphorylation of the mutated site. The most convincing experiments delete the wild type gene and replace it with the mutant version. These gene replacement experiments are easy to perform in yeasts but very difficult to do in all other organisms.

The effect of phosphorylating tyrosine 15 on Cdc2 activity was determined by replacing the fission yeast *cdc2⁺* gene with *cdc2-F15ᴰ,* a mutant version that replaced tyrosine 15 with a phenylalanine residue. The *cdc2-F15ᴰ* cells entered mitosis at a much smaller size than did wild-type cells (Figure 4-18) and could no longer arrest their cell cycles in response to drugs that inhibit DNA replication. These observations suggested that regulating tyrosine dephosphorylation of Cdc2 plays a critical role in regulating the progress of the fission yeast cell cycle engine. Cdc25 was no longer required for growth in *cdc2-F15ᴰ* strains. Strains of *cdc25ᵗˢ* arrest in G2 at 35°C with high levels of tyrosine 15 phosphorylation, strongly suggesting that the dephosphorylation of this site is required for entry into mitosis and that Cdc25 plays a key role in this reaction.

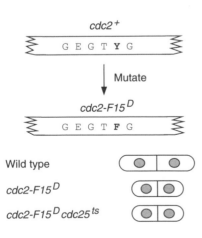

FIGURE 4-18 Regulation of mitosis by tyrosine phosphorylation of Cdc2. The top of the figure shows the amino acid sequence of Cdc2 around tyrosine 15 in wild-type Cdc2 and in the mutant where the tyrosine has been changed to phenylalanine (*cdc2-F15ᴰ*). The bottom part of the figure shows the cell sizes at septation of various mutants. Reference: Gould, K. L., & Nurse, P. *Nature* **342,** 39–45 (1989).

Wee1 catalyzes the phosphorylation and Cdc25 catalyzes the dephosphorylation of tyrosine 15 of Cdc2/28

Biochemical analysis of the purified proteins shows that Wee1 is a protein kinase that phosphorylates tyrosine 15 of Cdc2/28 and that Cdc25 is a protein phosphatase that removes the phosphate added by Wee1. In fission yeast, although Cdc25 and Wee1 are the major enzymes that regulate the tyrosine phosphorylation of Cdc2, other proteins exist that can perform the same functions. A gene called *mik1+* (mitotic inhibitory kinase) is closely related to *wee1+* and encodes a second kinase that can phosphorylate tyrosine 15 of Cdc2. The presence of a second kinase helps to explain why deleting *wee1+* disrupts the cell cycle much less than the complete block to phosphorylation caused by replacing tyrosine 15 with phenylalanine. Even in the absence of Wee1, Mik1 can catalyze some tyrosine phosphorylation. In accord with this explanation, *wee1ts mik1−* double mutants have a phenotype at least as severe as that of *cdc2-F15D* (Figure 4-19). Thus Mik1 and Wee1 are redundant proteins in the sense that one can partially substitute for the function of the other. Redundancy is the curse of genetics, which relies on identifying genes by the phenotypes that mutations in them produce. If two genes are fully redundant, single mutations in either one of them have no detectable phenotype. Even if the genes are only partially redundant, like *wee1+* and *mik1+*, the phenotype of the single mutations may be so weak as to be undetectable.

The activity of Cdc25 and Wee1 appears to be regulated during the cell cycle. The rate of phosphorylation and dephosphorylation of tyrosine 15 of Cdc2/28 has been measured in frog cell cycle extracts. In interphase tyrosine kinase activity is high and the phosphatase activity is low, but as the extracts approach mitosis the situation reverses: The phosphatase is activated and the kinase is inhibited. Direct measurements show that Cdc25 becomes phosphorylated and activated as extracts enter mitosis, suggesting that Cdc25 is the tyrosine phosphatase that acts on Cdc2/28 in crude extracts.

The phosphorylation of other sites on Cdc2/28 is much more poorly understood. Genetic analysis has not revealed the kinases and phosphatases that regulate the phosphorylation of threonine 161 in Cdc2/28 although biochemists have purified a candidate for a Cdc2/28-activating kinase. Phosphorylation of threonine 14, the residue adjacent to tyrosine 15, has been detected in mammalian cells. Like the phosphorylation of tyrosine 15, that of threonine 14 inhibits the kinase activity of Cdc2/28. The kinases and phosphatases that control the phosphorylation of this residue have not yet been identified.

The cyclin component of MPF also undergoes phosphorylations that change during the cell cycle, but their effect (if any) on MPF activity is not known. We also do not know how the association of cyclin with Cdc2/28 induces protein kinase activity. Does association with cyclin induce conformational changes in Cdc2/28 that favor protein kinase activation and/or does association with cyclin alter the pattern of post-translational modification on Cdc2/28 in a way that affects its activity? There is clear evidence that the latter alternative is at least partially correct. The phosphorylation of

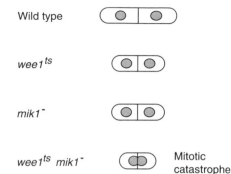

FIGURE 4-19 Effects of mutations in two tyrosine kinases that phosphorylate Cdc2. Wee1 and Mik1 are related tyrosine kinases that can both phosphorylate tyrosine 15 of Cdc2. The sizes at which cells of different strains septate are shown. The *wee1ts* mutant greatly reduces the size that a cell must reach to enter mitosis, while the *mik1−* mutant does not. The *wee1ts mik1−* double mutants cause cells to enter mitosis at such a small size that they undergo mitotic catastrophe and die. Reference: Lundgren, K., et al. *Cell* **64**, 1111–1122 (1991).

FIGURE 4-20 Cyclin degradation is triggered by ubiquitination. At the top of the figure the amino acid sequences of the amino-terminal regions of cyclins A and B1 are compared. The conserved region that is required for degradation is called the destruction box. Cyclin is targeted for degradation by conjugating ubiquitin molecules (Ub) to lysines downstream from the destruction box. The conjugates are recognized by the proteasome, a complex of proteolytic enzymes, only after several ubiquitins have been conjugated in a chain to each cyclin molecule. Reference: Glotzer, M., Murray, A. W., & Kirschner, M. W. *Nature* **349**, 132–138 (1991).

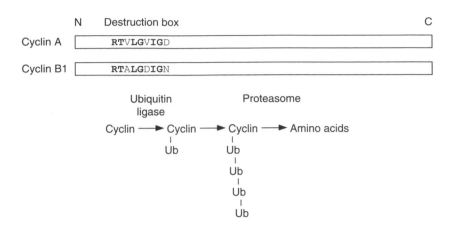

tyrosine 15 and threonine 161 normally requires association of Cdc2/28 with cyclin.

The recent determination of the three-dimensional structure of a member of the Cdc2 family has shown why monomeric Cdc2/28 lacks kinase activity. The part of the protein that includes threonine 161 occludes the substrate-binding site and distorts the active site of the protein. Presumably the combined effects of cyclin binding and threonine 161 phosphorylation cause a conformational change that overcomes these inhibitory effects.

The conjugation of multiple ubiquitin residues targets cyclin for degradation

The ability of nondegradable cyclins to arrest cells in mitosis demonstrates that cyclin degradation is required for the inactivation of MPF. The degradation of cyclin is the result of the conjugation of ubiquitin to cyclin (Figure 4-20). A complex series of reactions can conjugate a chain of ubiquitin residues to a single lysine in a protein, thus targeting the protein for degradation by a complex, cytosolic, multisubunit protease. In embryonic cells, cyclin ubiquitination is stimulated by MPF and therefore occurs only in mitosis. We do not know how cells regulate the ubiquitination of cyclin during the cell cyle or whether any steps in addition to cyclin degradation are required for the inactivation of MPF.

Mathematical models of the cell cycle engine can reproduce the key features of the embryonic cell cycle

An important test of our understanding of the cell cycle is our ability to create models that make testable predictions. The current biochemical model of the cell cycle engine is based on information from studies of many organisms (Figure 4-21). We begin at mitosis. Active MPF induces the degradation of cyclin. MPF may remain active as long as Cdc2/28 is phosphorylated on threonine 161, but in the absence of cyclin this phosphate is rapidly removed by phosphatases, inactivating MPF. Once MPF has been inactivated,

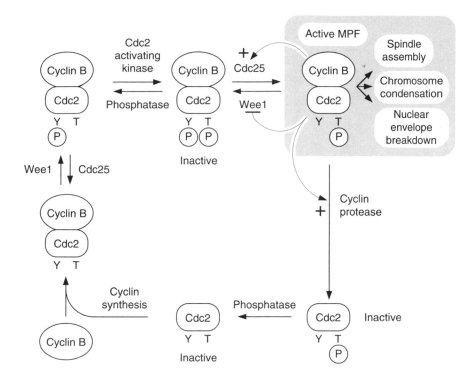

FIGURE 4-21 Biochemical model of the embryonic cell cycle engine. The different Cdc2/28–cyclin B complexes formed during the embryonic cell cycle and the enzymes that act on them are shown. The phosphorylation of Cdc2/28 on tyrosine 15 (left) and threonine 161 (right) is shown. At the beginning of interphase, Cdc2/28 is unphosphorylated. Association of cyclin with Cdc2/28 induces the phosphorylation of tyrosine 15 and threonine 161 to produce preMPF. At entry into mitosis, preMPF is converted into active MPF by Cdc25, which removes the phosphate from tyrosine 15. Active MPF induces the destruction of cyclin, resulting in MPF inactivation.

the activity of the cyclin degradation machinery decreases, so newly made cyclins A and B accumulate. As the cyclin accumulates, it binds to Cdc2/28, making Cdc2/28 a substrate for phosphorylation on tyrosine 15 by Wee1 and on threonine 161 by the Cdc2/28-activating kinase.

As long as tyrosine 15 is phosphorylated, the complex is in the preMPF form and lacks protein kinase activity. A small amount of active MPF is produced from molecules in which threonine 161 phosphorylation precedes that of tyrosine 15 or in which the tyrosine phosphate has been removed by a low level of Cdc25 activity. This active MPF is inactivated by the phosphatases that dephosphorylate threonine 161 or by Wee1, which rephosphorylates tyrosine 15.

When the pool of preMPF is small, the steady-state level of MPF is too low to turn on the reactions that will produce more MPF. But as cyclin accumulates, the pool of pre MPF swells, and the level of MPF increases until it eventually exceeds the ability of phosphatases and Wee1 to inactivate it. At this point MPF begins to phosphorylate and activate Cdc25 and to inhibit Wee1, dramatically increasing the rate of MPF activation and leading to a rapid and irreversible activation of all the preMPF in the cell. Once fully activated, MPF triggers the activation of the cyclin degradation machinery, leading to exit from mitosis. Although this model is doubtless incorrect in many details, it probably corresponds at least roughly to the actual events in the activation of MPF.

Mathematical models based on this formulation produce cell cycles in which the kinetics of cyclin accumulation and destruction and MPF activation

FIGURE 4-22 Mathematical model for the cell cycle. The top part of the diagram shows the biochemical pathways that have been incorporated into a mathematical model of the cell cycle engine. The small graph shows how the positive feedback on Cdc25 activity, the MPF-activating kinase, and cyclin degradation varies with the activity of MPF. Although the details of the biochemical pathways in this model differ slightly from those shown in Figure 4–21, these variations are not expected to have major effects on the kinetic behavior of the model. The lower part shows the fluctuation of cyclin and MPF levels predicted by the model for wild-type cells, a Δ*cdc25* mutant, a Δ*wee1* mutant, and a Δ*wee1* Δ*cdc25* mutant. The model reproduces the key features that removal of Cdc25 activity prevents entry into mitosis, the removal of Wee1 activity speeds up the cell cycle, and the removal of both activities generates an almost normal cycle. *Model constructed and simulations performed by Gary Odell.*

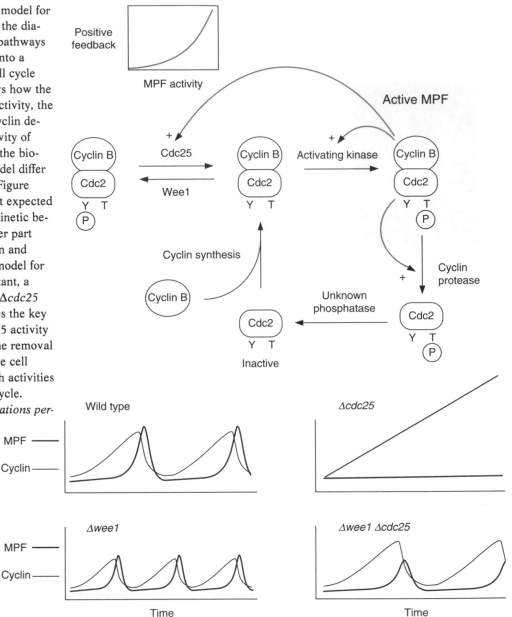

and inactivation roughly match the kinetics seen in vivo (Figure 4-22). To produce robust cycles, mathematical models need a cell cycle engine that shows strong positive feedback or that has delays built into it. The real cell cycle has both of these features. MPF is involved in two types of positive feedback. As cells enter mitosis, MPF stimulates its own activation by activating Cdc25 and inhibiting Wee1. Later, as cells prepare to leave mitosis, MPF stimulates its own inactivation by activating the proteolysis of cyclin. The degradation of cyclin at the end of mitosis and the need to resynthesize it before the next mitosis represents a built-in delay in the cell cycle, as does the

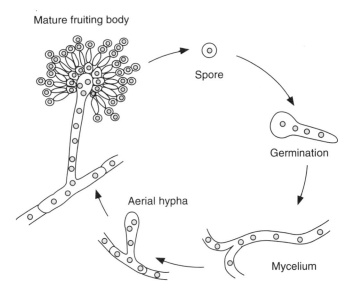

Mature fruiting body

Spore

Germination

Aerial hypha

Mycelium

FIGURE 4-23 *Aspergillus nidulans* life cycle. The asexual haploid stages of the *Aspergillus* life cycle are shown.

need to remove the tyrosine phosphate that prevents newly formed Cdc2/28 cyclin B complexes from acquiring MPF activity.

It is possible to construct models of cell cycle engines that use only post-translational modifications and do not include protein synthesis and degradation. Evolution may have selected protein synthesis and degradation partly because the protein synthesis requirement places an unavoidable delay in the cell cycle engine and partly because a regulated cycle of synthesis and degradation helps to enforce a one-way, irreversible cell cycle. We will return to this problem when we consider the mechanism of chromosome segregation (see pages 83–85).

MPF activity is not the only activity required to induce mitosis

Despite the encouraging successes of the last five years, our understanding of the events that drive cells into mitosis is incomplete. The idea that MPF is the sole inducer of mitosis has been challenged by experiments performed on the fungus *Aspergillus nidulans*. *Aspergillus* grows by forming long filaments, has a complicated life cycle (Figure 4-23), and like yeast, is well suited for genetic

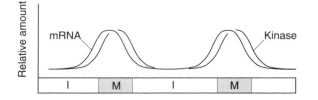

FIGURE 4-24 The NimA-associated kinase and mitosis. The level of NimA mRNA and the activity of NimA-associated kinase varies during the *Aspergillus nidulans* cell cycle. The mRNA increases just before mitosis, and the kinase activity peaks in mitosis. Reference: Osmani, A. H., O'Donnell, K., Pu, R. T., & Osmani, S. A. *EMBO J.* **10**, 2669–2679 (1991).

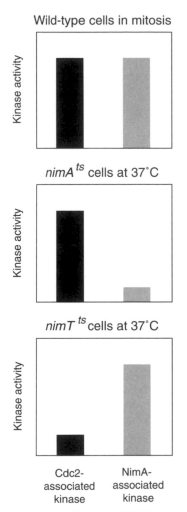

Wild-type cells in mitosis

nimA^{ts} cells at 37˚C

nimT^{ts} cells at 37˚C

FIGURE 4-25   NimA and Cdc2 act in parallel to induce mitosis. The activities of the NimA-associated and Cdc2-associated (MPF) protein kinases are shown in the conditions indicated. The *nimT*^{ts} mutant is a temperature-sensitive mutation in the *Aspergillus* homolog of the fission yeast *cdc25* gene. Reference: Osmani, A. H., McGuire, S. L., & Osmani, S. A. *Cell* **67**, 283–291 (1991).

analysis of the cell cycle. A large number of cell cycle mutants have been isolated in *Aspergillus*. These mutants are divided into two classes: *nim*^{ts} (never in mitosis) mutants, which cannot enter mitosis, and *bim*^{ts} (blocked in mitosis) mutants, which cannot exit mitosis.

Analysis of *nimA*^{ts} and *nimT*^{ts} mutants suggests that the activation of Cdc2/28 is not sufficient to induce entry into mitosis. The *nimA*⁺ gene encodes a protein kinase, and the levels of both NimA mRNA and NimA-associated protein kinase activity increase dramatically as cells enter mitosis (Figure 4-24). Strong overexpression of NimA can induce premature entry into mitosis, even in cells that have not yet completed DNA replication. As expected, when *nimA*^{ts} mutants are arrested at the restrictive temperature, they lack NimA-associated protein kinase activity. But these same cells contain high levels of MPF, showing that, although the activation of MPF is independent of NimA activity, it is not sufficient to induce entry into mitosis.

Analysis of mutations in *nimT*⁺, which is a homolog of the fission yeast *cdc25*⁺ gene, was equally revealing (Figure 4-25). Since Cdc25 activity is required for the activation of MPF in fission yeast, it is no surprise that *nimT*^{ts} *Aspergillus* has little MPF activity when arrested at the restrictive temperature. However, these cells do have high levels of NimA-associated kinase activity, suggesting that the activation of the NimA-associated kinase does not require MPF activity and is not sufficient to induce mitosis. Thus in *Aspergillus*, MPF and the NimA-associated kinase are both required for entry into mitosis and seem to be activated independently of each other.

CONCLUSION

This chapter shows how information from research on several different organisms converged to reveal that the cell cycle engine in all eukaryotes is fundamentally similar. During evolution the components of the engine have been so strongly conserved that many of them can be swapped between organisms without seriously disrupting the operation of the cell cycle. The cell cycle is regulated at two checkpoints—Start and entry into mitosis—at which signals generated by a variety of inputs, including cell size, nutrient availability, signals from other cells, and the completion of downstream events, can arrest progress of the cell cycle. Much of the variation among different cell cycles reflects the extent to which they respond to these signals. The early cell cycles of the frog embryo are highly simplified and fail to respond to any signals, while those of somatic cells are tightly regulated and respond to all of them.

SELECTED READINGS

Reviews

Forsburg, S. L., & Nurse, P. Cell cycle regulation in the yeasts *Saccharomyces cervisiae* and *Schizosaccharomyces pombe*. *Ann. Rev. Cell Biol.* **7**, 227–256 (1991). *A review of the cell cycle in fission and budding yeast.*

Draetta, G., ed. Cyclin dependent kinases. *Seminars in Cell Biol.* **2**, 193–271 (1991). *A series of reviews on the biochemistry and biology of Cdc2/28–cyclin complexes.*

Hunt, T., & Kirschner, M. W., eds. Cell multiplication. *Curr. Opinion in Cell Biol.* **5**, Issue 2 (1993). *A series of reviews on various aspects of the cell cycle. This publication reviews advances in the cell cycle each year.*

Original Articles

Nurse, P. Genetic control of cell size at cell division in yeast. *Nature* **256**, 547–551 (1975). *Identification of threshold sizes for mitosis and Start.*

Beach, D., Durkacz, B., & Nurse, P. Functionally homologous cell cycle control genes in budding and fission yeast. *Nature* **300**, 706–709 (1982). *Demonstration that fission yeast cdc2$^+$ is equivalent to budding yeast CDC28$^+$.*

Lohka, M. J., Hayes, M. K., & Maller, J. L. Purification of maturation-promoting factor, an intracellular regulator of early mitotic events. *Proc. Natl. Acad. Sci. USA* **85**, 3009–3013 (1988). *Purification of MPF.*

Murray, A. W., & Kirschner, M. W. Cyclin synthesis drives the early embryonic cell cycle. *Nature* **339**, 275–280 (1989). *Use of cell cycle extracts to show the role of cyclin synthesis in inducing mitosis.*

Gould, K. L., & Nurse, P. Tyrosine phosphorylation of the fission yeast cdc2$^+$ protein kinase regulates entry into mitosis. *Nature* **342**, 39–45 (1989). *Demonstration that tyrosine phosphorylation of Cdc2 regulates the activity of MPF.*

5   MITOSIS

*H*OW DOES active MPF induce cells to enter mitosis and how does the mitotic spindle segregate the cell's chromosomes? Chromosome movement and cell division have fascinated biologists for more than a hundred years. Until the 1970s the main approach was descriptive. Scientists observed mitotic cells first by light microscopy and later by electron microscopy. Although their detailed observations of normal cells and of cells treated with various drugs led to a plethora of models for mitosis, they produced little reliable information about the molecular structure or function of the spindle.

Since 1970 great advances in our understanding of the proteins that make up the spindle have been made, and students of mitosis now stand at the brink of a revolution much like the one that has transformed our understanding of the cell cycle. Our new knowledge of mitosis comes from advances in biochemistry, light microscopy, and genetics. Biochemical analysis of microtubules and the proteins that interact with them led to the crucial revelation that the spindle is a tremendously dynamic structure. The rapid remodeling of the spindle during mitosis reflects the ability of microtubules to grow and shrink quickly, as well as the existence of **microtubule motors** that can move along microtubules in the same way that myosin can move along actin filaments. A renaissance in light microscopy has led to a remarkable improvement in our ability to observe living cells and the behavior of specific molecules in them.

Light microscopy   **Microtubule staining**

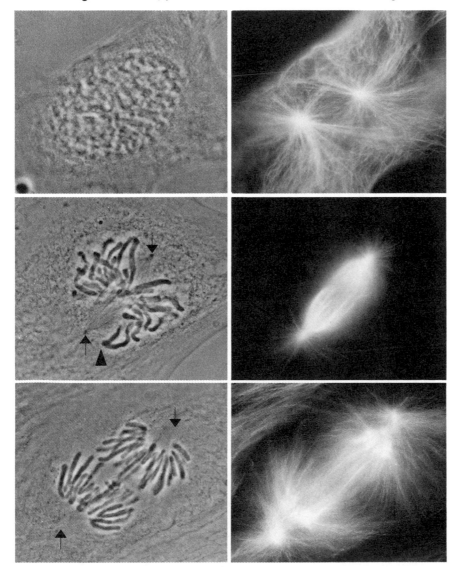

FIGURE 5-1 Mitosis: a series of light micrographs showing the progression of mitosis in a newt lung cell. The left-hand column shows the appearance of the cell seen by phase contrast microscopy, while the right-hand column shows the pattern of microtubules inside the cell as visualized by staining with fluorescent antitubulin antibodies. From top to bottom, the stages are prophase, metaphase and anaphase. In prophase the centrosomes have already separated from each other, although they are still nucleating the long microtubules characteristic of interphase cells. In metaphase almost all the microtubules are located within the spindle and all but one of the chromosomes have moved to points equidistant from the two poles (the kinetochore of the exceptional chromosome is indicated with an arrowhead). In anaphase the sister chromatids have separated from each other and moved toward the poles (anaphase A), the distance between the centrosomes has increased (anaphase B), and new arrays of cytoplasmic microtubules have begun to appear. The positions of the centrosomes are indicated with arrows in the metaphase and anaphase light micrographs. Photographs courtesy of Bob Skibbens and Ted Salmon.

Finally, genetics has helped to identify components of the spindle and their role inside cells. In particular, genetics identified the **centromeres,** the DNA sequences on chromosomes that combine with a number of proteins to form the **kinetochore,** the specialized structure that attaches the chromosomes to microtubules.

Like the cell cycle as a whole, mitosis in different cells can look very different, but the fundamental mechanisms that assemble spindles and segregate chromosomes appear to be strongly conserved. The most obvious difference is between open mitosis and closed mitosis (see Figure 1-6). Most of this chapter describes open mitosis, whose morphological stages were discussed in Chapter 1.

Figure 5-1 shows the profound changes in the organization of the chromosomes and cytoskeleton that occur as a mammalian cell goes through mitosis. Spindle assembly requires chromosome condensation, nuclear envelope breakdown, and changes in the properties of microtubules. MPF appears to induce these processes independently of each other, rather than directing a single coordinated program of events that produces the mitotic spindle. We will therefore discuss the separate processes individually before trying to understand how they interact to produce the process of mitosis and how these interactions are regulated to ensure faithful chromosome segregation.

NUCLEAR ENVELOPE BREAKDOWN

The key feature of open mitosis is the disappearance of the nucleus, which occurs during the later stages of chromosome condensation. The nuclear envelope is composed of an inner and outer nuclear membrane and a proteinaceous shell or **lamina** that lies on the nuclear side of the inner membrane (Figure 5-2). The two membranes lie close together with the outer surface of the outer membrane in contact with the cytoplasm. The narrow space between the membranes is continuous with the lumen of the endoplasmic reticulum. The nuclear membranes are punctuated by **nuclear pores,** complicated structures that are anchored in the lamina. The pores span the inner and outer nuclear membranes and allow the exchange of components between the nucleus and cytoplasm.

Phosphorylation of nuclear lamins induces their disassembly

The lamina is a meshwork of filaments that are polymers of the **nuclear lamins,** which are members of the class of proteins that form the intermediate filaments that give cells much of their rigidity (Figure 5-3). Most cells contain two principal lamins, lamin A and lamin B. When the nuclear envelope breaks down, the filaments of the nuclear lamina dissolve, the nuclear mem-

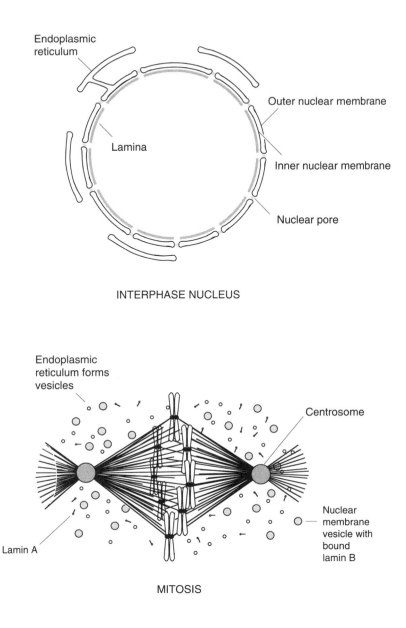

FIGURE 5-2 Nuclear structure. In interphase the double membrane that forms the nuclear envelope lies on top of a proteinaceous lamina composed of the polymerized nuclear lamins. The lumen between the two nuclear membranes is continuous with the endoplasmic reticulum. Nuclear pores are embedded in the nuclear envelope and allow communication between nucleus and cytoplasm. As cells enter mitosis the nuclear envelope vesiculates, the nuclear lamins depolymerize, and the nuclear pores break down. After depolymerization, lamin A is free in solution, but lamin B remains attached to membrane vesicles.

brane is converted into small vesicles (see Figure 5-2), and the pores break down into smaller subassemblies.

The depolymerization of the filaments that make up the lamina is induced by phosphorylation of lamin A and B as cells enter mitosis. Phosphorylation of lamin A converts it into lamin monomers free in solution, whereas phosphorylated lamin B monomers remain associated with the membrane vesicles because of a hydrophobic isoprenyl group at their C-terminus that inserts them into the lipid bilayer of the membranes (see page 150). When the nuclear envelope reassembles at the start of interphase, lamin B may form sites that initiate reassembly of the lamina.

Mutant lamins that cannot be phosphorylated do not depolymerize in mitosis. The human lamin A gene was modified by mutating the principal

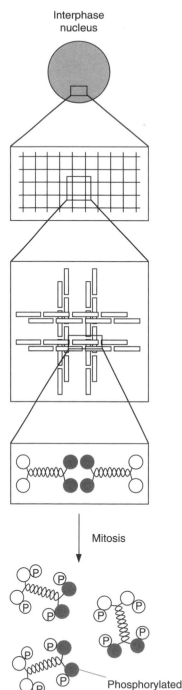

Interphase
nucleus

Mitosis

Phosphorylated
lamin A dimer

FIGURE 5-3 Nuclear lamina structure. In interphase, lamin molecules form dimers that associate head to head to form a tetrameric unit that is the repeating structure of the polymerized lamin filaments. The lamina is a meshwork of these lamin filaments. In mitosis, the phosphorylation of lamin molecules at the end of their rod-shaped domains induces the disassembly of the lamin filaments and the breakdown of the nuclear lamina. References: Moir, R. D., Donaldson, A. D., & Stewart, M. *J. Cell Sci.* **99**, 363–372 (1991); Aebi, U., Cohn, J., Buhle, L. & Gerace, L. *Nature* **323**, 560–564 (1986).

mitotic phosphorylation sites to nonphosphorylatable alanine residues. The mutated human lamin gene was introduced into hamster cells. When the cells entered mitosis, the hamster lamins depolymerized but the human ones did not (Figure 5-4), although the nuclear membrane dispersed normally and a spindle formed with microtubules passing through holes in the lamina. In addition to showing that phosphorylation is needed to dissolve the lamina, this experiment showed that neither nuclear membrane fragmentation nor spindle assembly is dependent on complete dissolution of the lamina.

The phosphorylated serines in the lamins flank the rodlike region of lamin A and match the consensus sequence for sites phosphorylated by MPF (a serine or threonine followed by a proline accompanied by nearby basic amino acids). When lamin filaments are incubated with purified MPF and ATP, they depolymerize. Based on this evidence we suspect that the lamins are phosphorylated in vivo by MPF. Unfortunately, there is no easy or general method for showing that a protein is phosphorylated directly by MPF rather than by a kinase that MPF activates.

As they disassemble, the nuclear membranes are converted from sheets of membranes into small vesicles. The Golgi apparatus and endoplasmic reticulum are also composed of sheets of membrane that are converted into vesicles during mitosis. In interphase, traffic between the different membrane compartments involves vesicles that bud from one compartment and then fuse with another. Adding MPF to interphase frog cell cycle extracts induces the vesiculation of the nuclear membranes, endoplasmic reticulum, and Golgi apparatus, probably by suppressing vesicle fusion. We do not know the molecular basis of these changes in membrane organization or whether MPF acts directly or by activating other protein kinases.

Nuclear re-formation occurs by fusion of vesicles bound to chromosomes

The re-formation of the nuclear envelope has been studied in frog egg extracts. Decondensing chromosomes seem to have receptors on their surface that bind membrane vesicles, and these vesicles fuse with each other to form a continuous double membrane around each chromosome (Figure 5-5). The nature of the proteins on the vesicles and of the decondensing chromosomes that form the critical initial interactions in nuclear re-formation is unknown.

In embryonic cell cycles the chromosomes decondense when they reach the spindle poles to form individual mininuclei or **karyomeres,** which fuse with each other and form a single nucleus that contains all the chromosomes.

The lamins associate with the karyomeres only after the nuclear membrane has re-formed; they are probably imported via the nuclear pores and reassemble into the characteristic meshwork once inside. In cell cycle extracts, lamins will only polymerize inside nuclei. Perhaps the presence of membrane associated lamin molecules is required to initiate lamin polymerization.

The extent to which nuclear membrane, nuclear pore, and nuclear lamin assembly depend on each other is a controversial subject. Lamin assembly apparently requires nuclear membranes, but nuclear membrane assembly can occur in the absence of free lamins or pore assembly. Nuclear pore assembly is also independent of lamina assembly. But although a nuclear membrane can assemble without a lamina or nuclear pores, these components are critical for nuclear function. Without lamins, nuclei are extremely fragile and incapable of DNA replication, and without pores, they cannot import the cytoplasmic proteins required for replication or transcription.

CHROMATIN CONDENSATION

Eukaryotic chromosomes must condense before they are segregated. The average human chromosome contains about 130 million base pairs, or a 45 mm length, of DNA, whereas the mitotic spindle is only 10 to 20 μm long. Metaphase chromosomes condense until they are only a few microns long, representing a 10,000-fold contraction of the length of the naked DNA they contain. Condensation makes it possible for the chromosomes to move without becoming entangled and probably also makes them strong enough to withstand mechanical forces that could otherwise break their DNA.

Changes in chromatin condensation occur throughout the cell cycle

The first sign that a typical animal or plant cell is about to divide is that its chromosomes start to condense visibly. Chromosome condensation can be seen in living cells by phase-contrast microscopy or with fluorescent DNA-

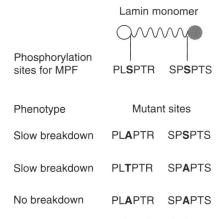

FIGURE 5-4 Lamin phosphorylation is required for lamina breakdown. The top of the figure shows a schematized lamin molecule with globular N- and C-terminal domains and a rod-like central region. The amino acid sequence of the sites where MPF phosphorylates the lamins are shown. Converting either of these sites — serine (S) or threonine (T) — to an alanine residue (A) reduces the rate of lamina disassembly; converting both sites to alanine blocks lamina disassembly completely. Reference: Heald, R., & McKeon, F. *Cell* **61**, 579–589 (1990).

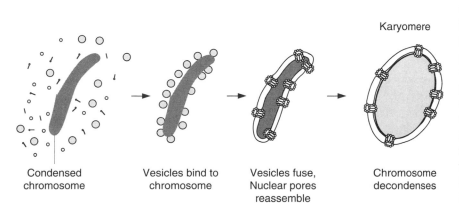

FIGURE 5-5 Pathway of nuclear assembly. Nuclear assembly involves the progression from a condensed chromosome to a mininucleus (karyomere) at the end of mitosis. First, membrane vesicles bind to the chromosome in a reaction that requires chromatin associated proteins and soluble factors. Then the vesicles fuse, and once fusion is complete, chromosome decondensation occurs. In embryonic cell cycles, DNA replication begins as the chromosomes are decondensing and while the karyomeres are fusing to create a single interphase nuclei. Reference: Vigers, G. P., & Lohka, M. J. *J. Cell Biol.* **112**, 545–556 (1991).

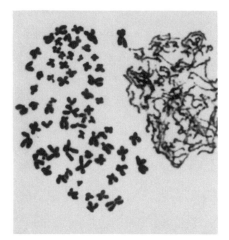

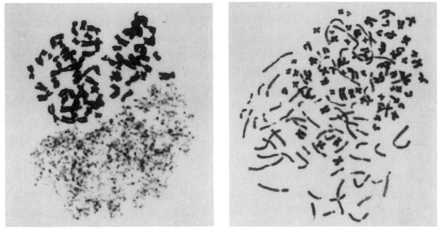

FIGURE 5-6 Chromosome condensation during the cell cycle. The appearance of chromosomes in hybrid cells produced by fusing cells in different stages of the cell cycle to cells in mitosis is shown. In each micrograph the chromosomes from the mitotic cells are highly condensed and the two sister chromatids are distinct. Chromosomes derived from G1 cells (left panel) are very thin and extended, those from S phase cells (middle panel) are apparently broken into many tiny fragments, and those from G2 cells (right panel) show paired sister chromatids but are much longer than those from mitotic cells. Reference: Johnson, R. T., & Rao, P. N. *Biological Reviews* **46**, 97–155 (1971). Micrographs courtesy of W. N. Hittleman.

binding dyes. During interphase the DNA appears as a homogeneous fluorescent cloud, which condenses into a granular tangle of threads at the start of prophase. As mitosis progresses, the threads get shorter and fatter. In open mitosis, chromosome condensation typically begins early in prophase, so individual chromosomes can already be seen when the nuclear envelope breaks down, although chromosomes often become even more condensed as the cell proceeds to metaphase. At the end of mitosis, as the chromosomes reach the spindle poles, they begin to decondense and, in open mitoses, acquire a nuclear envelope.

Although changes in chromosome condensation are obvious as cells enter mitosis, cell fusion experiments reveal that the state of the chromosomes changes throughout the cell cycle (Figure 5-6). When G1 cells are fused to mitotic cells, the chromosomes from the G1 nucleus form threads that are very long compared to the compact pairs of sister chromatids from the mitotic cell. S phase nuclei in mitotic cytoplasm produce clouds of microscopic chromosome fragments, suggesting that replicating chromosomes cannot be condensed properly. When G2 cells are fused with mitotic cells, the chromosomes from the G2 cell form visible sister chromatids but are still considerably less condensed than those from the mitotic cell, suggesting that condensation occurs continuously between the end of S phase and the onset of mitosis. In embryonic cells that have been arrested in interphase by blocking protein synthesis, considerable chromosome condensation occurs within the intact nucleus even in the absence of detectable MPF activity.

Chromosome condensation is poorly understood

We are surprisingly ignorant of the molecular basis of chromosome condensation and decondensation. These events are among the first that students of mitosis identified and were meticulously described long before the central role of DNA in chromosome structure and function was known. The biochemistry of histones and other chromosome-associated proteins has been studied intensively, but the inability to reproduce chromosome condensation in extracts was a major stumbling block. The ability of cell cycle extracts

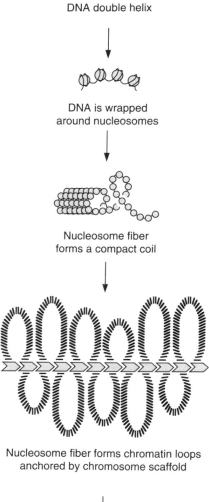

FIGURE 5-7 Levels of chromosome condensation. The diagram shows the different levels of compaction involved in packaging DNA into chromosomes. The basis of chromosome structure is the formation of nucleosomes by wrapping naked DNA around a core of 8 histone molecules (top). The nucleosomal fiber is further compacted by winding it into a solenoidal filament 30 nm in diameter. This filament is them arranged into loops (that contain about 50,000 base pairs of DNA) that are attached to the chromosome scaffold. Each chromosome is made up of many such loops (bottom).

prepared from frog eggs to perform chromosome condensation should assist our attempts to understand the mechanism of this important process.

Chromosomes are enormous macromolecular assemblies whose major components are DNA and **histones,** although they contain many other proteins. Histones are small, basic proteins that associate to form a ball of eight histone molecules that has about 160 base pairs of DNA wound around it in two helical turns to create a structure called a **nucleosome.** These globular structures are arranged along the DNA like beads on a string. The formation of nucleosomes reduces the physical length of a piece of DNA sevenfold. The compaction of mitotic chromosomes is a thousand times greater than this. Some of the compaction results when the beads are wound on a string into a solenoidal fiber and then this fiber is arranged as a series of loops that average about 50,000 base pairs in length (Figure 5-7). These loops are thought to be the basic units of chromosome transcription and replication and are supported by the **chromosome scaffold,** which is defined as the residue of chromosomal proteins that remains after all the histones and most of the DNA have been removed from the chromosome.

Are changes in chromatin condensation due to changes in the kind or number of proteins bound to chromatin or to modifications of proteins that remain bound to chromatin throughout the cell cycle? Which proteins are involved in such changes? The answers to these basic questions are unknown. Even for the histones, where changes in protein modification during the cell cycle have been extensively analyzed, we do not know how these changes relate to chromosome condensation. Histone H1, which is bound to the DNA between nucleosomes, becomes progressively phosphorylated on serine and threonine as cells enter mitosis and is an excellent substrate for MPF. Although the sites of phosphorylation have been identified (Figure 5-8), the biological consequences of phosphorylating them are not known.

Topoisomerase II is an important chromosomal protein. This enzyme can untangle DNA by breaking one DNA duplex, passing another duplex through the gap, and then closing the gap. Changes in the topology of chromosomal DNA seem to be important for chromosome condensation, since inhibitors of type II DNA topoisomerases block chromosome condensation in mitotic frog egg extracts. Perhaps these results indicate that the tangles must be unraveled as chromosome condensation proceeds. Topoisomerase II inhibitors do not block nuclear envelope breakdown or spindle assembly, which shows that chromosome condensation is not a prerequisite for later events of mitosis.

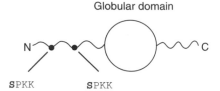

Globular domain

N ⌇•⌇•⌇ C

SPKK   SPKK

Nuclear lamins
LSPTR
PSPTS

Lamin B receptor
 TPEK

Histone H1
 TPAK
 SPKK
 TPKK

pp60 c-src
QTPNKT
RTPSRS
SSPQRA

Nucleolin
 TPAK
 TPGK
 SPKK

SV40 T-antigen
 TPPKK

RNA polymerase
 YTPTSPSY

Myosin light chain
RATSNVF (!)

Rab-4
 SPRR

Cyclin A
ASPMVV

Cyclin B
PSPVPM

FIGURE 5-8 MPF-induced phosphorylation sites. This diagram shows the positions at which MPF phosphorylates histone H1. The sites that MPF phosphorylates in a number of other proteins are also shown. Amino acid sequences are indicated in the one-letter code. Reference: Nigg, E. A. *Seminars in Cell Biol.* **2**, 261–270 (1991).

MICROTUBULES, MICROTUBULE MOTORS, AND KINETOCHORES

The formation of the mitotic spindle that will segregate the chromosomes depends on the activity of MPF. The spindle is an architecturally complex and surprisingly dynamic structure, whose shape reflects a balance between opposing forces. Altering this balance produces rapid changes in the structure of the spindle. This section describes the properties of microtubules, microtubule motors, and kinetochores that lead to the formation of the spindle.

Microtubules are dynamic, polar polymers

Microtubules are the major structural component of the spindle and were first seen by electron microscopists in the 1950s. Microtubules are built from dimers of α and β tubulin. Each microtubule consists of 13 individual strings of tubulin dimers, and these polymers are aligned side by side to form a hollow tube 25 nm in diameter (Figure 5-9). Individual tubulin subunits are asymmetric, and as a result microtubules are polar structures with distinct plus and minus ends that have different properties, such as the rates at which microtubules grow and shrink. Many different **microtubule-associated proteins (MAPs)** bind to microtubules. Some of these proteins are structural proteins, which bind to the sides or ends of microtubules and affect their stability, while others are microtubule motors that can move along the surface of microtubules by hydrolyzing ATP. Like microtubules, the motors are polar: One class moves toward the plus end of a microtubule, the other toward the minus end. Cellular structures can be attached to motor proteins and moved along microtubules.

The mitotic spindle can undergo rapid structural changes. In the most extreme case, the spindle of the early *Drosophila* embryo forms, segregates the chromosomes, and disassembles in only four minutes. The rapid assembly and disassembly of the mitotic spindle is made possible by the rapid turnover of the microtubules themselves. The kinetic behavior of microtubules is called **dynamic instability** (Figure 5-10). Microtubules are not stable structures. They exist in one of two different dynamic states, growing or shrinking. Cytoplasm contains a pool of free tubulin dimers, so for microtubule ends in the growing state the rate of tubulin addition greatly exceeds the rate of tubulin loss. When in the shrinking state, however, the structure of the microtubule ends is different, and the rate of subunit loss far exceeds that of subunit addition. Microtubule ends undergo abrupt (and unpredictable) tran-

FIGURE 5-9   Microtubule structure. Microtubules are nucleated at and radiate from the centrosome, which consists of a pair of centrioles and an associated cloud of proteins. Microtubules are repeating polymers of a dimer composed of one α and one β tubulin molecule. The polymers are hollow tubes 25 nm in diameter, with 13 tubulin subunits arranged around the circumference of the tube. The orientation of the a tubulin dimer within the polymer is not known.

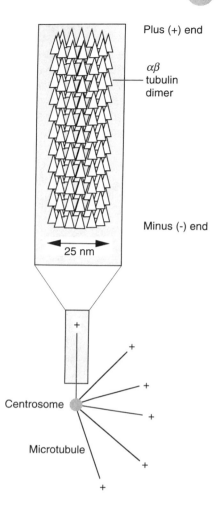

Plus (+) end

αβ
tubulin
dimer

Minus (-) end

25 nm

Centrosome

Microtubule

sitions from the growing state to the shrinking state (**catastrophe**) or from the state shrinking to the growing state (**rescue**).

The number and length of the microtubules in a cell are a function of the concentration of tubulin dimers and the rates of microtubule nucleation, polymerization, depolymerization, catastrophe, and rescue. Changes in any one of these parameters, either temporally or spatially within the cell, can change the form and distribution of microtubules. Some microtubule-associated proteins bind to the surface of microtubules and alter the catastrophe and rescue frequencies as well as accelerating the rate of polymerization. In some structures, such as cilia, flagella, and nerve axons, microtubule-associated proteins can stabilize microtubules for long periods.

MPF induces rapid microtubule turnover

In living cells, microtubule assembly is nucleated from microtubule organizing centers. A typical interphase animal cell contains a single active microtubule organizing center, the centrosome, which lies close to the nuclear envelope and nucleates long microtubules that run out in all direction towards the periphery of the cell. This organized array is called an **aster**. The average lifetime of an astral microtubule is about 5 minutes (see Figure 5-1).

An important component of the microtubule organizing center is **γ tubulin**. This molecule is related to the α and β tubulin that make up the subunits of microtubules, but γ tubulin is restricted to regions around the microtubule organizing centers, where it appears to play an important role in nucleating microtubule polymerization.

As cells approach mitosis and MPF activity increases, the microtubule array in the cell undergoes a dramatic reorganization. The average length of the microtubules decreases about tenfold, and the number of microtubules nucleated by each centrosome increases tenfold. Studies on the dynamics of individual microtubules in interphase and mitotic extracts have shown that the change in microtubule length is due to the ability of active MPF to induce a marked increase in the catastrophe frequency and a decrease in the rescue frequency. MPF probably alters microtubule dynamics by inducing the phosphorylation of microtubule associated proteins that control microtubule dynamics. What are these proteins and are they phosphorylated directly by MPF or by kinases activated by MPF? The answers to these questions are unknown.

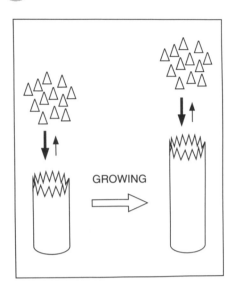

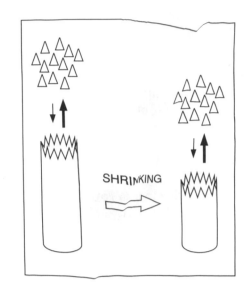

FIGURE 5-10 Dynamic instability of microtubules. Microtubules exist in one of two states, growing or shrinking. In the growing state subunits add to the plus ends of the microtubules much faster than they leave them, but in the shrinking state subunit loss far exceeds addition. The conversion from a growing to a shrinking state is called catastrophe, while that from a shrinking to a growing state is called rescue. In mitosis the catastrophe frequency increases and the rescue frequency decreases, resulting in a large number of short, unstable microtubules that form the spindle. References: Mitchison, T., & Kirschner, M. *Nature* **312**, 237–242 (1984); Belmont, L. D., Hyman, A. A., Sawin, K. E., & Mitchison, T. J. *Cell* **62**, 579–589 (1990).

Microtubule motors are crucial components of kinetochores

Microtubule motors are mechanochemical enzymes that use the energy derived from ATP hydrolysis to move along the surface of the microtubules. The known motors belong to two families: **dyneins** and **kinesins** (Figure 5-11). The dyneins are very large, complex proteins composed of multiple subunits. All known dynein-like motors move rapidly toward the minus ends of microtubules. The kinesins are smaller proteins that are active as dimers. Some members of the kinesin family move toward the plus ends of microtubules; others move toward the minus ends. As far as we know, no motor is capable of moving in both directions. It is clear that motors can move objects along microtubules, but they may have an equally important function as dynamic anchors of objects to microtubules.

Genetic and biochemical studies in budding yeast have defined the centromeric DNA and identified kinetochore proteins that bind to it. The kinetochores of budding yeast are extremely small, with each chromosome interacting with a single microtubule. The centromeric DNA in budding yeast occupies only 130 base pairs and is composed of three distinct sequence elements, of which the most important is about 25 base pairs long (Figure 5-12). This element is absolutely required for proper centromere function and acts as the binding site for a complex of three proteins. When the purified protein complex is incubated with centromeric DNA, it forms a complex that binds to microtubules and can use the energy from ATP hydrolysis to move toward their minus ends. These experiments mark the beginning of the reconstruction of a functional kinetochore in vitro and promise to yield valuable information about its structure and function.

Studies on the centromeres and kinetochore proteins of other organisms are much less advanced. In fission yeast, a stretch of DNA more than 10,000 base pairs long contains the functional centromeric DNA. We do not know

FIGURE 5-11 Microtubule motors. The figure shows three microtubule motors. Dynein moves towards the minus ends of microtubules. Kinesin-like motors are divided into two subclasses. One group, which includes kinesin itself, moves toward the plus end of the microtubules. The other, which includes Ncd (a motor involved in the assembly of the meiotic spindle in flies), moves towards the minus ends of microtubules.

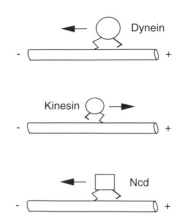

why the centromeric DNA of fission yeast is so much larger than that of budding yeast, since the kinetochores of both organisms bind only one microtubule. Perhaps the greater extent of chromosome condensation in fission yeast necessitates complicated sequences that flank the kinetochore and prevent it from becoming buried as the cells enter mitosis.

The kinetochores of vertebrate cells are complex, forming hemispherical protrusions, or kinetochore plates, that extend from the surface of the chromosome. Although repeated DNA sequences (called α-satellite DNA) that are associated with the kinetochore regions of human chromosomes have been isolated, the functional centromeric DNA of vertebrates has not been identified. Some kinetochore proteins have been identified using antibodies from patients suffering from scleroderma, an autoimmune disease in which the body makes antibodies to its own proteins. A small fraction of scleroderma patients have antisera that recognize components of the kinetochore. Four different proteins that interact with such sera have been identified, including a protein that binds to α-satellite DNA, a histonelike protein, and a form of kinesin. Antibodies to known proteins have also identified components of the kinetochore by showing that mitotic chromosomes contain a form of dynein at their kinetochores.

What purpose do microtubule motors at the kinetochore serve? An obvious answer is that motor proteins could explain how chromosomes move during mitosis, but motors are probably also important for tethering chromosomes to microtubules. A tour de force of light microscopy revealed that the interactions of microtubules with centrosomes and kinetochores are dynamic. The experiment took advantage of small molecules that become fluorescent only when they have been irradiated with ultraviolet light. These potentially fluorescent tags were chemically coupled to tubulin dimers that were injected into living metaphase cells. The cells were then illuminated with a narrow bar of ultraviolet light running across the long axis of the spindle so that the fluorescence of the injected tubulin was activated as a discrete mark on the kinetochore microtubules. The movement of the fluorescent

FIGURE 5-12 Structure of the budding yeast kinetochore. The DNA sequence of the budding yeast centromere occupies about 130 base pairs and is divided into three functional elements. Of these, the most important is element III, a conserved DNA sequence found at the centromere of every yeast chromosome, which acts as a binding site for a complex of three kinetochore proteins that play a vital role in attaching the chromosome to a microtubule. References: Fitzgerald-Hayes, M., Clarke, L., & Carbon, J. *Cell* **29**, 235–244 (1982); Hyman, A. A., Middleton, K., Centola, M., Mitchison, T. J., & Carbon, J. *Nature* **359**, 533–536 (1992).

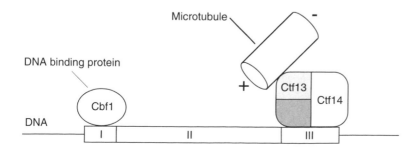

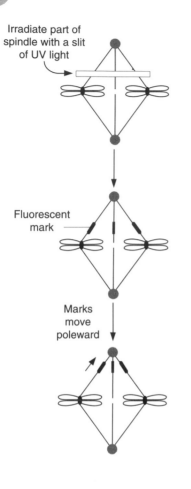

Irradiate part of spindle with a slit of UV light

Fluorescent mark

Marks move poleward

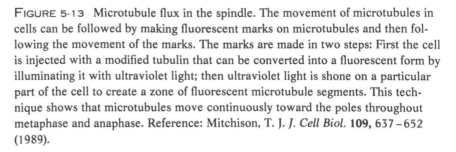

FIGURE 5-13 Microtubule flux in the spindle. The movement of microtubules in cells can be followed by making fluorescent marks on microtubules and then following the movement of the marks. The marks are made in two steps: First the cell is injected with a modified tubulin that can be converted into a fluorescent form by illuminating it with ultraviolet light; then ultraviolet light is shone on a particular part of the cell to create a zone of fluorescent microtubule segments. This technique shows that microtubules move continuously toward the poles throughout metaphase and anaphase. Reference: Mitchison, T. J. *J. Cell Biol.* **109,** 637–652 (1989).

mark was followed using a very sensitive video camera (Figure 5-13). During metaphase the separation between the pole and the chromosomes remains constant, but the marks on the kinetochore microtubules move slowly and steadily toward the spindle poles. This movement demonstrates that in metaphase microtubules lose subunits at the centrosome and gain them at the kinetochore: Although the length of the microtubule is constant, the centrosome is attached to a shrinking end and the kinetochore to a growing end.

How can a motor anchor the pole or kinetochore to a microtubule whose end is growing or shrinking? The most likely explanation requires that a microtubule end abuts the motor-containing structure. For instance, the dynein-like motors at the kinetochore could hold on to the shrinking plus end of a microtubule. As long as the motor moved toward the minus end faster than subunits were lost from the plus end of the microtubule, the motor would always remain attached to the shrinking microtubule (Figure 5-14). According to this model, the chromosome can move toward the pole only as fast as the microtubule depolymerizes, so the rate of chromosome movement is governed by the rate of microtubule depolymerization rather than the intrinsic speed of the motor.

FIGURE 5-14 Microtubule motors as anchors. Microtubule motors can hold an object to a shrinking microtubule as long as the motor can move along the microtubule faster than the wave of depolymerization.

A polar wind repels objects from the spindle poles

The microtubule arrays nucleated by the spindle poles repel chromo and other large particles, such as mitochondria or yolk granules. This revealed graphically by using a laser to perform surgery on the spin

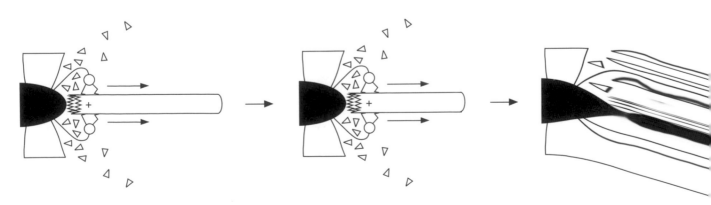

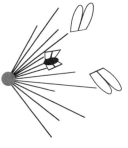

Chromosome
arms severed
with laser beam

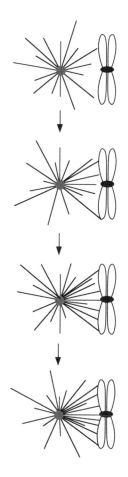

FIGURE 5-15 Polar wind repels objects from the spindle poles. Microtubules growing out from the spindle poles push objects away from the poles. If the arms of a chromosome that is attached to the pole are cut off, the severed arms are pushed away from the pole, while the central fragment of the chromosome that bears the kinetochore moves toward the pole. Reference: Rieder, C. L., Davison, E. A., Jensen, L. C. W., Cassimeris, L., & Salmon, E. D. *J. Cell Biol.* **103,** 581–591 (1986).

living cell. If both arms of a chromosome are severed from its kinetochore, the free arms are expelled from the spindle, while the fragment that carries the kinetochore moves closer to the spindle pole. The force that repels objects from the spindle poles is called the **polar wind** or **astral ejection force** (Figure 5-15). The nature of the wind is not clear. It might simply result from growing microtubules physically pushing objects away from the pole, like a knight being repelled by a lance. Alternatively, plus-end directed motors attached to the chromosome arms (and the other organelles, too) might generate a force directed toward the plus end of the microtubules.

Kinetochores and chromatin can protect microtubules from catastrophe

What accounts for the asymmetric distribution of microtubules in the spindle? An isolated centrosome nucleates a symmetrical array of microtubules, but in many metaphase spindles the vast majority of the microtubules are directed toward the center of the spindle. A cross section through the metaphase plate reveals that the center of the spindle is a dense array of microtubules free of chromosomes, while the chromosomes are attached to microtubules around the edge of the spindle. This distribution probably reflects differential stabilization of microtubules in different parts of the spindle. Even if the spindle pole nucleates microtubules equally in all directions, the interaction of some of them with structures that protect them from catastrophe can produce a very asymmetric distribution. In mitotic cells microtubules can be stabilized in at least three ways: by being captured by kinetochores, by forming tight arrays with microtubules of opposite polarity, or by interacting with chromatin (Figure 5-16). We do not understand the molecular basis of stabilization in any of these cases.

FIGURE 5-16 Chromatin stabilizes microtubules. Microtubule distribution can be affected by selective stabilization of some microtubules. Microtubules are dynamic: They grow away from and shrink back toward the centrosome. In the absence of other factors, the microtubules are spherically distributed around the centrosome. Microtubules that interact with chromatin, however, are stabilized—that is, their catastrophe frequency is reduced—so the distribution of microtubules becomes asymmetric. Reference: Karsenti, E., Newport, J., Hubble, R., & Kirschner, M. *J. Cell Biol.* **98,** 1730–1745 (1984).

SPINDLE ASSEMBLY

Unlike the structure of DNA, which immediately suggests that a preexisting strand acts as the template for a new daughter strand, the structure of the mitotic spindle does not suggest how it is established. There is no hint of a template on which the spindle can be built. We are forced to the conclusion that it is built according to rather general principles and achieves its final form through processes that involve a good deal of trial and error. In the next section we review what is known about the processes that must occur to produce a bipolar spindle (a spindle with two distinct poles), on which the chromosomes are aligned on a metaphase plate with the kinetochores of the two sister chromatids attached to opposite poles.

An early visible step of mitosis is the separation of the centrosomes. In many cells centrosome separation precedes nuclear envelope breakdown, although in some, it follows nuclear envelope breakdown. Injecting antibodies that inhibit the function of cytoplasmic dynein in animal cells prevents centrosome separation. In fission yeast and *Aspergillus,* mutations that inactivate members of the kinesin family prevent the separation of spindle pole bodies. This evidence suggests a crucial role for microtubule motors in catalyzing the separation of the spindle poles.

Microtubule motors enable kinetochores to capture microtubules

When the nuclear envelope breaks down, the kinetochores begin to interact with the polar microtubules. One of the best places to study these early interactions is in newt lung cells, which are very thin, are optically clear, and have large chromosomes that make them excellent specimens for light microscopy. When these cells enter mitosis, some chromosomes lie far from the poles in regions of low microtubule density, making the sequence of events involved in chromosome attachment to the poles easy to see (Figure 5-17). The kinetochore plate initially binds to the side (rather than to the end) of a single microtubule. Once attached, the chromosome moves rapidly toward the pole at about 50 μm per minute, a speed that is characteristic of dynein motors. As the chromosome nears the pole, it slows down and eventually stops, attaining a steady-state position at which it oscillates toward and away from the pole. At this stage multiple microtubules that terminate in the kinetochore plate attach the chromosome to the spindle pole. Collectively these microtubules comprise a **kinetochore fiber** that terminates at the kinetochore facing the pole; the other kinetochore, which faces away from the pole, has no associated microtubules. The steady-state position of chromosomes attached to a single pole probably reflects a balance of forces, with the polar wind pushing the chromosome away from the pole and the dynein-like molecules at the kinetochore pulling the kinetochore toward the pole.

Eventually, the unattached kinetochore interacts with a microtubule from the other pole, developing a second kinetochore fiber and causing the

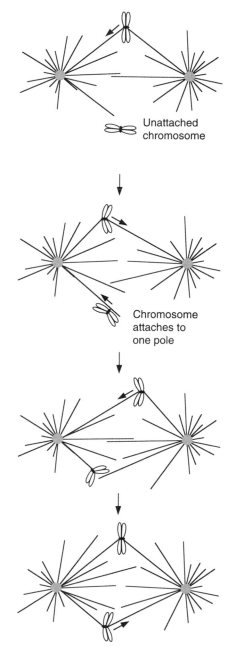

FIGURE 5-17   The diagram shows four successive views of a spindle in a cell approaching metaphase. At the top of the diagram one pair of sister chromatids is attached to both poles of the spindle, while the other is free in the cytoplasm. In the next view one kinetochore of the previously unattached pair has bound a single microtubule and motors at the kinetochore are moving the chromosome polewards; the other chromosome, which has moved closer to one spindle pole, has just reversed direction and is now moving toward the other pole. In the third view the pair of sisters that has just attached to the spindle has established a monopolar orientation, in which multiple microtubules are bound to (and end at) the kinetochore facing the pole, while the other kinetochore is devoid of microtubules. The chromosome whose attachment is bipolar has moved from one side of the metaphase plate to the other as a result of the periodic reversals in the direction in which the kinetochores of an individual chromosome move. In the final view the sister chromatid pair whose attachment was monopolar has captured microtubules from the opposite spindle pole to establish a bipolar orientation and is moving towards the metaphase plate. Reference: Rieder, C. L., & Alexander, S. P. *J. Cell Biol.* **110**, 81–95 (1990).

Unattached chromosome

Chromosome attaches to one pole

chromosome to move towards the center of the spindle. The details of the balance of forces on the chromosomes and poles at metaphase remain to be worked out. Motors at the kinetochores that pull them toward the spindle poles tend to pull the poles together, since the two sister kinetochores are connected to each other. To prevent the spindle from shrinking, this force must be opposed by one or more forces that keep the spindle poles apart. There are two candidates for opposing force: motors in the middle of the spindle, where microtubules from opposite poles overlap, that would tend to push the poles apart (essentially a form of polar wind) and minus-end directed motors attached to the cytoskeleton or cell cortex that would pull the microtubules that point away from the metaphase plate toward the periphery of the cell (Figure 5-18). Evidence for each of these forces exists in different cells, and the relative importance of the different forces may well differ in different cell types.

The existence of different mechanisms for maintaining the length of the metaphase spindle suggests that, as in the cell cycle engine itself, there is considerable redundancy in the spindle. For many of the individual problems that the spindle has to solve, more than one solution exists. Like the components of the cell cycle engine, it is likely that most of these mechanisms are strongly conserved in evolution, but the relative importance of different mechanisms in different cell types can vary widely, giving the appearance that there are many different ways of doing mitosis.

What ensures that the two sister kinetochores attach to opposite poles of the spindle? One simple but important factor is that the kinetochores of a pair of sister chromosomes face in opposite directions; thus when one kinetochore attaches to one pole, its sister faces the opposite pole. Nevertheless, in prometaphase both kinetochores occasionally attach to the same pole. If this state persisted to anaphase, it would result in both sisters segregating to one daughter cell. We do not understand how errors in chromosome orientation are rectified in mitotic cells.

FIGURE 5-18 Forces on the metaphase spindle. In metaphase, kinetochore motors that exert force toward the spindle poles pull the spindle poles toward each other. This force is opposed either by interaction of microtubules with the cortex of the cell (upper diagram) or by interactions between microtubules of opposite polarities in the center of the spindle (lower diagram).

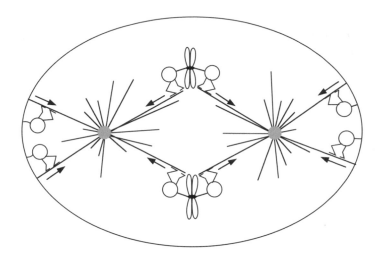

Kinetochore forces balanced by astral forces

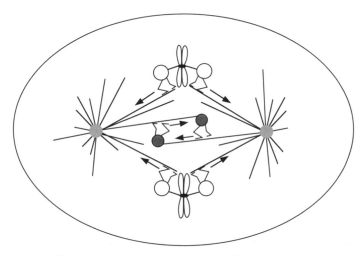

Kinetochore forces opposed by spindle overlap forces

Chromosomes congress to the middle of the spindle

As cells approach metaphase, the chromosomes **congress:** They align on the metaphase plate, equidistant from the spindle poles. The process that aligns the chromosomes on the plate is called **congression.** To produce a steady-state position equidistant from the two poles, some force acting on the chromosomes must vary in intensity as the distance of the chromosome from the pole changes. The two forces whose distance dependence has been proposed to account for congression are the polar wind and the force that kinetochores exert toward the pole.

At metaphase, the chromosomes are not stationary on the spindle; instead, individual chromosomes oscillate about an average position, moving first toward one pole and then reversing direction and moving toward the opposite pole (see Figure 5-17). Whatever the origin of this oscillation is, its existence emphasizes the dynamic nature of the spindle.

ANAPHASE

The proteolysis of cyclin and the inactivation of MPF trigger two classes of downstream events. One group of events is simply the reversal of events that occur when cells enter mitosis. Chromosome decondensation, nuclear envelope re-formation, and changes in microtubule dynamics are probably the consequence of removing phosphate groups that were added to proteins as cells entered mitosis. This idea is supported by the fact that mutations in type 1 phosphatases in fission yeast, *Aspergillus,* and *Drosophila* all slow down or completely block exit from mitosis. Type 1 phosphatases may play a role in the inactivation of MPF, as well as in the removal of phosphate groups that were added to proteins by active MPF.

Anaphase and cytokinesis represent a second class of events that are induced by the inactivation of MPF. Neither of these events is a simple reversal of a change that occurred as cells entered mitosis. For anaphase to occur, certain precursor events, such as formation of a mitotic spindle, must be induced by the activation of MPF. But as long as cyclin is stable and MPF is active, anaphase cannot occur.

Cyclin protease triggers the separation of sister chromatids

Accurate chromosome separation requires that sister chromatids be held together from the time of DNA replication until anaphase, when the linkage between sisters must be promptly dissolved, allowing the two sisters to segregate to opposite poles of the spindle. What triggers sister chromatid separation, how do the chromosomes move to the spindle pole, and what holds sisters together? We know a good deal about the first question, something about the second, and very little about the third. Although the separation of sister chromatids normally becomes visible as they are pulled toward the spindle poles in anaphase, separation does not require microtubules pulling on kinetochores. In cells that contain chromosome fragments lacking kinetochores, the sister chromatids of the fragments show an initial separation from each other at the same time as the normal chromosomes.

Much of our information on the trigger of anaphase comes from studies on frog cell cycle extracts. Using extracts prepared from unfertilized eggs that have high levels of MPF, it is possible to form metaphase spindles in vitro. Like the meiosis II spindle in an unfertilized egg, these spindles remain stably arrested in metaphase until an increase in the intracellular calcium level induces the degradation of cyclin and the inactivation of MPF. In vivo, fertilization induces the rise in calcium; in extracts, calcium is simply added to the reaction.

In normal cells the activation of the cyclin degradation machinery, the degradation of cyclin, and the inactivation of MPF follow inevitably one after the other. The availability of nondegradable forms of cyclin allow experimenters to separate the activation of cyclin degradation from the inactivation of MPF (see page 54). When extracts containing nondegradable cyclin are treated with calcium, the endogenous cyclin is degraded, but MPF activity remains high because of the presence of the nondegradable cyclin. In these extracts, the sister chromatids separate from each other and move close to the poles of the spindle, although the spindle does not break down and the chromosomes remain condensed (Figure 5-19). This experiment demonstrates that sister separation does not require the inactivation of MPF or the degradation of all the cyclin in the extract.

If sister chromatid separation does not require the degradation of cyclin, does it depend on the proteolytic machinery that degrades cyclin? The existence of peptides derived from the N-terminus of cyclin that can act as competitive inhibitors of cyclin degradation makes it possible to answer this

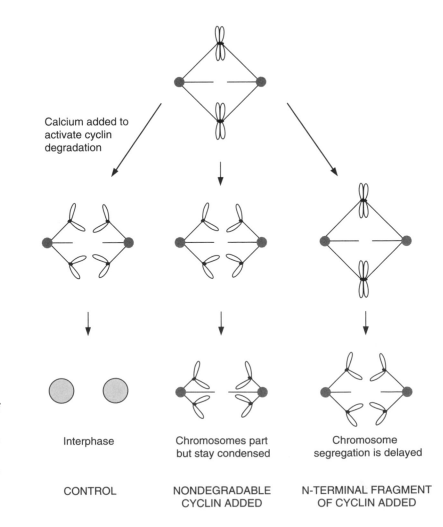

FIGURE 5-19 Proteolysis induces chromosome segregation. Spindles can be assembled in extracts of frog eggs and induced to enter anaphase by adding calcium to the extracts. In control extracts, the addition of calcium induces the degradation of cyclin and the inactivation of MPF just before sister chromatid separation occurs. In extracts that contain a nondegradable form of cyclin, however, the addition of calcium induces sister chromatid separation but does not lead to the inactivation of MPF. In extracts that contain an N-terminal fragment of cyclin, the inactivation of MPF and sister chromatid separation are delayed.

Calcium added to activate cyclin degradation

Interphase

Chromosomes part but stay condensed

Chromosome segregation is delayed

CONTROL

NONDEGRADABLE CYCLIN ADDED

N-TERMINAL FRAGMENT OF CYCLIN ADDED

question. When calcium is added to a metaphase extract containing the inhibitory peptides, degradation of the endogenous cyclin, sister chromatid separation, and MPF inactivation are all delayed by the same amount.

Thus, two experiments give apparently paradoxical results: experiments with nondegradable cyclin suggest that cyclin degradation is not required for sister separation, but experiments with the inhibitory peptide suggest that the proteolytic activity that degrades cyclin is required to separate the sisters. The simplest explanation of these observations is that there is a protein, other than cyclin, that must be degraded by the cyclin protease in order to break the linkage between sister chromatids. Degrading a protein cross-link between the sisters may represent a particularly effective way of holding sisters tightly together during interphase and metaphase and then rapidly freeing them from each other in anaphase.

Separation of sister chromatids requires DNA topoisomerase II

One of the linkages between sister chromatids is a topological linkage that is an inevitable consequence of the mechanism of DNA replication. The final stage of DNA replication produces daughter molecules that are linked to each other or **catenated** (Figure 5-20). Consider what happens at the termination of replication, when two replication forks traveling in opposite directions meet. During DNA replication the two strands of the DNA duplex must unwind from each other, and topoisomerases act as swivels to keep this unwinding from generating a hopeless tangle. As the forks meet, the topoisomerases are squeezed out, so the last part of replication occurs without the benefit of the swivel. As a result, each turn of the helix that is replicated generates one intertwining of the daughter molecules. Since eukaryotic chromosomes have multiple replication origins, there will be multiple points of linkage between the sisters. Although the chromosomes are linear, the attachment of the DNA to a chromosome scaffold at intervals of about 50,000 base pairs (page 73) means that, topologically, each chromosome can be thought of as a chain of circles, and at metaphase each circle is catenated to the corresponding circle on its sister. To separate the sisters, the catenation of sister chromatids must be resolved by passing one DNA double helix through the other, the reaction catalyzed by type II DNA topoisomerases.

Experiments on several organisms show that sister chromosomes are catenated to each other at metaphase. Temperature-sensitive mutants in type II topoisomerases of budding and fission yeast initiate anaphase but are unable to separate their sister chromosomes. Even in frog spindles that have been assembled in vitro and arrested at metaphase for long periods, the action of topoisomerase II is still required at the time of anaphase, demonstrating that decatenation can occur at anaphase but not at metaphase. Total topoisomerase II activity shows no significant increase when MPF is inactivated, suggesting that the DNA is inaccessible to topoisomerases during metaphase. In addition, some other linkage besides DNA catenation may hold the sisters together. Possible candidates for proteins that hold sisters together include

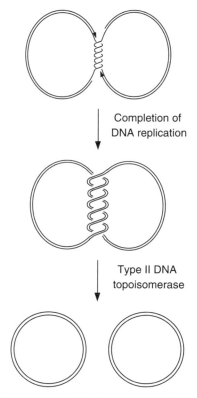

Completion of
DNA replication

Type II DNA
topoisomerase

FIGURE 5-20 Topoisomerase II decatenates sister chromatids. During the last stages of DNA replication, chromosomes replicate without unwinding, leaving the daughter DNA duplexes wound around one another. This linkage is resolved by the action of type II DNA topoisomerases, which transiently cut one DNA duplex and pass the other one through the gap.

the INCENP proteins (inner centromere proteins), which were identified by human autoimmune sera that stain the region between sisters at metaphase. At anaphase these antigens are left behind by the segregating chromosomes at the site where the metaphase plate was. Although the localization of these proteins is very suggestive, there is no direct evidence that they play a role in holding sisters together.

Sister chromatid segregation triggers anaphase chromosome movement

The dissolution of the linkage between sisters may be the cause of, as well as a prerequisite for, the characteristic movements of chromosomes and poles at anaphase. The metaphase spindle is, as we have seen, in a state of tension; microtubule motors at the kinetochores exert a force on the kinetochores that tends to pull the poles together, but the spindle is kept from collapsing by opposing forces.

Once the linkage between sister chromosomes is broken, the polewards forces on the kinetochores are no longer in opposition, and the chromosomes are free to move toward the poles. Thus, without any change in the magnitude or direction of the forces acting on the spindle, the separation of the sisters can bring about anaphase chromosome movement (Figure 5-21). Estimates of the poleward force at the kinetochore suggest that it does not change during the transition from metaphase to anaphase.

In animal cells the movement of the chromosomes to the poles is due both to kinetochores following the depolymerizing ends of the kinetochore microtubules and to the poleward transport of the microtubules that is seen even in metaphase. In the early part of anaphase, about two-thirds of the movement is due to microtubule depolymerization at the kinetochore, and the remaining third is due to the microtubule being reeled in at the centrosome. In the latter part of anaphase, depolymerization of the kinetochore end of the microtubule ceases, and the final movement toward the pole is due entirely to the loss of subunits at the centrosomal end of the microtubule. In many organisms, the density of astral microtubules decreases during anaphase, leading to a decrease in the force of the polar wind. This, too, facilitates movement of the chromosomes towards the poles.

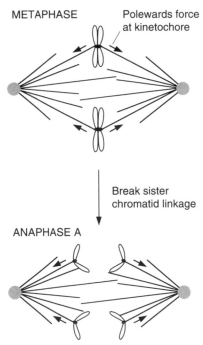

METAPHASE   Polewards force at kinetochore

Break sister chromatid linkage

ANAPHASE A

FIGURE 5-21 Sister chromatid separation triggers anaphase. Metaphase is a dynamic steady state maintained by the balance between forces pulling the chromosomes toward the spindle poles and forces pushing the spindle poles away from each other. Breaking the linkage between the chromosomes destroys the linkage between these forces and allows the chromosomes to move toward the spindle poles (anaphase A).

Inactivation of MPF triggers cytokinesis

The final cytoskeletal event triggered by the inactivation of MPF is cytokinesis. In animal cells cytokinesis is due to the formation of a ring containing actin and myosin just under the cortex of the cell. The contraction of this ring acts like a purse string pinching the cell apart into two daughter cells. What localizes the contractile ring, and what activates the contractile machinery after the onset of anaphase?

The site at which the contractile ring forms is clearly specified by the position of the spindle: The ring forms where the cortex is intersected by a

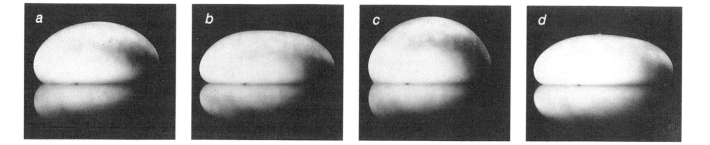

plane that passes through the metaphase plate. Once anaphase begins, the spindle can be physically removed from the cell without affecting cleavage, showing that the role of the spindle is restricted to specifying the cleavage plane. The mechanism by which the spindle specifies the cleavage plane is unknown.

The activation of the contractile machinery is triggered by the inactivation of MPF. Frog eggs that have been activated rather than fertilized cannot form a spindle, since they lack the centrosome that is normally contributed by the sperm. These eggs do not cleave; instead, the inactivation of MPF triggers a wave of contraction that spreads out over the surface of the egg (Figure 5-22). This observation demonstrates that although the spindle is required to localize contractile activity in the form of a cleavage furrow, the activation of the contractile machinery is independent of the spindle. Agents that block the inactivation of MPF, such as nondegradable cyclins, prevent the contraction wave, strongly suggesting that it is dependent on the inactivation of MPF.

One clue to how MPF inactivation may trigger contraction comes from studies of myosin phosphorylation. The light chain of myosin can be phosphorylated on two types of sites. Phosphorylation at certain positions activates the ATPase activity of myosin and increases its affinity for actin fila-

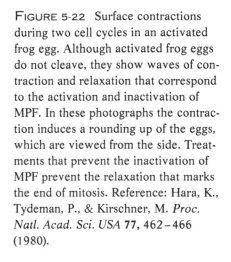

FIGURE 5-22 Surface contractions during two cell cycles in an activated frog egg. Although activated frog eggs do not cleave, they show waves of contraction and relaxation that correspond to the activation and inactivation of MPF. In these photographs the contraction induces a rounding up of the eggs, which are viewed from the side. Treatments that prevent the inactivation of MPF prevent the relaxation that marks the end of mitosis. Reference: Hara, K., Tydeman, P., & Kirschner, M. *Proc. Natl. Acad. Sci. USA* **77**, 462–466 (1980).

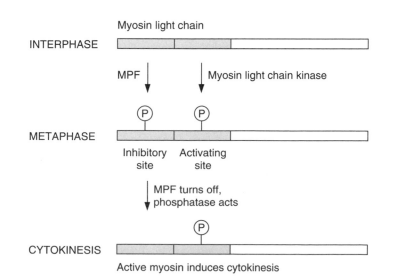

FIGURE 5-23 Myosin dephosphorylation triggers cytokinesis. MPF induces the phosphorylation of the light chain of myosin, preventing it from binding to actin and hydrolyzing ATP. These changes prevent cells from attempting to cleave before the end of mitosis. The inactivation of MPF at the end of mitosis leads to the dephosphorylation of myosin, the activation of the contractile machinery, and cytokinesis. Reference: Satterwhite, L. L., et al. *J. Cell Biol.* **118**, 595–605 (1992).

ments. Phosphorylation at other positions decreases the ATPase activity of myosin, lowers its affinity for actin, and inhibits phosphorylation on the activating site (Figure 5-23). Purified MPF can phosphorylate the inhibitory sites in vitro, suggesting that active MPF inhibits contraction and explaining why MPF must be inactivated before the embryo can cleave.

Making chromosome segregation dependent on proteolysis, while cytokinesis is dependent on the inactivation of MPF, may help to ensure that chromosome segregation occurs before cytokinesis. Sister separation would be triggered at the same time as cyclin degradation, but cytokinesis could not begin until any other steps required to inactivate MPF had occurred and the MPF-induced phosphorylations that block cytokinesis had been reversed.

CONCLUSION

MPF initiates a number of processes as cells enter mitosis. Apart from nuclear lamina dissolution, we do not understand the molecular mechanisms of any of the changes induced by MPF. In the case of chromosome condensation, we do not even understand the general principles that alter the architecture of the chromosomes. The general rules of spindle assembly have been identified, and considerable progress has been made toward identifying the molecules that form the kinetochores and spindle poles and modulate the behavior of microtubules. During the next few years we should see rapid progress toward understanding the molecular details of spindle assembly, chromosome movement, and sister chromatid separation and toward discovering how MPF produces the dramatic structural rearrangements that occur as cells go through mitosis.

SELECTED READINGS

Reviews

Nicklas, R. B. Mitosis. *Adv. Cell Biol.* **2**, 225–97 (1971). *An excellent review of the older literature on mitosis.*

Mitchison, T. J. Microtubule dynamics and kinetochore function in mitosis. *Ann. Rev. Cell Biol.* **4**, 527–549 (1988). *A discussion of the conceptual issues in spindle assembly.*

McIntosh, J. R., & Hering, G. E. Spindle fiber action and chromosome movement. *Ann. Rev. Cell Biol.* **7**, 403–426 (1991).

Rieder, C. L. Mitosis: Towards a molecular understanding of chromosome behavior. *Curr. Opinion Cell Biol.* **3**, 59–66 (1991). *Two recent reviews of recent work on spindle assembly and function.*

Original Articles

Heald, R., & McKeon, F. Mutations of phosphorylation sites in lamin A that prevent nuclear lamina disassembly in

mitosis. *Cell* **61**, 579–589 (1990). *Phosphorylation of nuclear lamins is required for disassembly of the lamina.*

Belmont, L. D., Hyman, A. A., Sawin, K. E., & Mitchison, T. J. Real-time visualization of cell cycle-dependent changes in microtubule dynamics in cytoplasmic extracts. *Cell* **62**, 579–589 (1990). *Microtubules become less stable upon the activation of MPF.*

Mitchison, T. J. Polewards microtubule flux in the mitotic spindle: Evidence from photoactivation of fluorescence. *J. Cell Biol.* **109**, 637–652 (1989). *Microtubules can move relative the kinetochore and spindle pole without becoming detached.*

Rieder, C. L., & Alexander, S. P. Kinetochores are transported poleward along a single astral microtubule during chromosome attachment to the spindle in newt lung cells. *J. Cell Biol.* **110**, 81–95 (1990). *Chromosomes can attach to and move along a single microtubule.*

THE TRANSITION FROM G1 TO S PHASE

6

SOMATIC CELL cycles vary in length, unlike those of early embryos, whose duration is rigidly programmed. In early embryos the inactivation of MPF triggers the events that lead to DNA replication and duplication of the microtubule organizing center. By contrast, growing cells have a transition in G1 that regulates the initiation of DNA replication and the duplication of the microtubule organizing center. This transition, called Start in unicellular eukaryotes and the restriction point in vertebrates, accounts for much of the flexibility in somatic cell cycles.

In budding yeast and mammalian tissue culture cells, Start is the major checkpoint that coordinates progress of the cell cycle engine with cell growth, nutrient availability and signals from other cells. What molecules induce Start? How is Start regulated? Our current understanding of Start relies heavily on analogies drawn from the studies of MPF described in Chapter 4. The observations that Cdc2/28 plays a crucial role in the induction of Start and mitosis and that MPF is made up of Cdc2/28 and mitotic cyclins suggested that there were G1 cyclins that combine with Cdc2/28 to induce Start. We shall see that although the discovery of G1 cyclins in budding yeast verified this prediction, their existence was revealed by investigations of how cell size and mating factors regulate Start, rather than by a direct search for new cyclins.

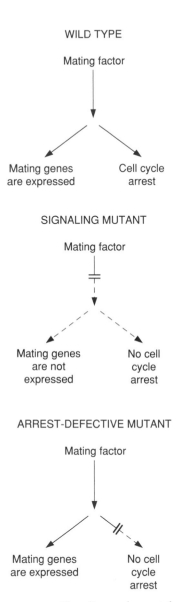

WILD TYPE

Mating factor

Mating genes are expressed

Cell cycle arrest

SIGNALING MUTANT

Mating factor

Mating genes are not expressed

No cell cycle arrest

ARREST-DEFECTIVE MUTANT

Mating factor

Mating genes are expressed

No cell cycle arrest

FIGURE 6-1   Signaling and arrest-defective mutants. In wild-type cells mating pheromones induce two pathways: expression of mating genes and cell cycle arrest. Many signaling mutants prevent mating pheromones from inducing either of these responses. Arrest-defective mutants, which can express mating genes but cannot arrest their cell cycle, identify genes that are important in controlling the cell cycle engine.

An important question is how similar the regulation of Start in yeast and the control of the restriction point in vertebrate cells are. The first half of this chapter describes in detail the regulation of Start in yeast; the second half assesses our much scantier knowledge about the restriction point in mammalian cells. Although the details of cell cycle control clearly differ, there are many striking similarities between the enzymes that control Start and those that control the restriction point.

START CONTROL IN BUDDING YEAST

The budding yeast *cdc28^{ts}* mutant keeps cells from passing through Start at the restrictive temperature. Analysis of this mutant showed that cell size, nutrient availability, and mating factors regulate Start (see Chapter 3). Do these very different factors all regulate the activity of Cdc28 through the same component of the cell cycle engine? To answer this question, it was necessary to find mutations that alter the regulation of Start rather than simply blocking this transition.

Mating-factor-resistant mutations identify a signaling pathway that regulates Start

Because cells can grow without mating, it is possible to isolate nonconditional mutations that alter a cell's response to mating factors. Chemically synthesized α factor prevents **a** cells from passing through Start, providing a powerful tool for selecting mutations that affect the regulation of Start; only those mutant **a** cells that no longer respond to α factor can give rise to colonies on a medium that contains α factor.

When mating factors bind to their receptors, they activate a complex signaling pathway. Although each receptor can bind only one type of mating factor, the structures of the **a**-factor and α-factor receptors are similar, and the subsequent steps in the pathway are identical in the two cell types. At its end the pathway splits into two branches: One induces mating genes that are involved in the physical act of conjugation; the other arrests the cell cycle at Start. The mating-factor-resistant mutants fall into two classes: signaling mutants and arrest-defective mutants (Figure 6-1). **Signaling mutants** block steps before the split in the mating-factor-response pathway. These mutants behave as if they cannot detect the factor: Mating factors do not induce mating genes or cell cycle arrest. **Arrest-defective mutants** block steps after the split in the response pathway; thus, they can induce mating genes but not cell cycle arrest. These mutants can still mate, although they do so at a reduced frequency.

Studies of signaling and arrest-defective mutants in budding yeast have made two crucial contributions to our understanding of the cell cycle. An arrest-defective mutant identified the G1 cyclins that activate the Start inducing activity of Cdc28, and the combined analysis of the two classes of mutants produced the best molecular description of a response pathway that leads from the cell surface to the entrails of the cell cycle engine (pages 93–96).

A dominant α-factor-resistant mutant identifies a G1 cyclin

A simple scheme that explains how mating factors arrest the cell cycle also suggests that mating factors activate a pathway that leads to destruction of one of the components needed to induce Start. In this model, two types of mutation prevent mating factors from arresting the cell cycle: recessive mutations that inactivate components of the signaling pathway and mutations that alter components of the engine so that it is no longer sensitive to the signaling pathway.

A selection for dominant mating-factor-resistant mutants repeatedly isolated mutations in a budding yeast gene called *CLN3* (cyclin). The *CLN3^D* mutants had two phenotypes: Mating factors no longer arrested the cell cycle, and even in the absence of mating factors, the mutant cells were smaller than wild-type cells (Figure 6-2). Other *CLN3^D* mutations were isolated in a screen for small cells, much like the one that had produced the *wee1^ts* mutations in fission yeast. Since Start is the main point at which size regulates the budding yeast cell cycle, the reduced size of *CLN3^D* mutants suggested that this gene might control Start, in much the same way that the analysis of dominant *cdc2-w* mutants in the fission yeast had revealed the role of Cdc2 in inducing mitosis (see pages 44–46). Changing the dosage of the wild-type *CLN3* gene confirmed that Cln3 plays an important role in regulating Start. Cells with multiple copies of the *CLN3* gene passed Start at a smaller size than wild-type cells, but cells whose *CLN3* gene had been deleted had to become larger than wild-type cells before they could pass Start.

Cln3 is weakly homologous to mitotic cyclins and is called a G1 cyclin because of its role in inducing Start. The *CLN3^D* mutants produce a truncated protein that is much more stable than wild-type Cln3 protein. The sequences required for the degradation of mitotic cyclins are located at their amino termini, whereas the stability of the Cln3 mutants, which lack the carboxyl third of the wild-type protein, suggests that those of G1 cyclins are at the carboxyl termini (see Figure 6-2).

Three redundant G1 cyclins control start

At the same time that *CLN3* was discovered, a completely different approach identified two other genes, *CLN1* and *CLN2*, that showed weak but detectable homology to mitotic cyclins. *CLN1* and *CLN2* are much more closely related to each other than they are to *CLN3*. In much the same way that either of the two frog B-type cyclins is sufficient to induce mitosis, expression of any one of the three *CLN* genes will allow budding yeast cells to pass Start.

Like the mitotic cyclins, the G1 cyclins must be made afresh in each cell cycle. This conclusion was reached by constructing budding yeast strains that

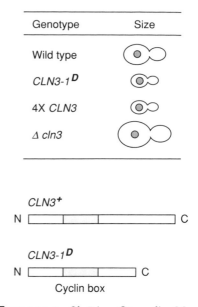

FIGURE 6-2 Cln3 is a G1 cyclin. Mutations in the *CLN3* gene affect cell size and sensitivity to α factor. The lower part of the figure shows the structure of the Cln3 protein and the carboxyl-terminal deletion that increases the stability of the protein. The cyclin box is the region of the protein that is homologous to other cyclins. Reference: Cross, F. R. *Mol. Cell. Biol.* **8,** 4675–4684 (1988).

CELLS GROWN WITH GALACTOSE

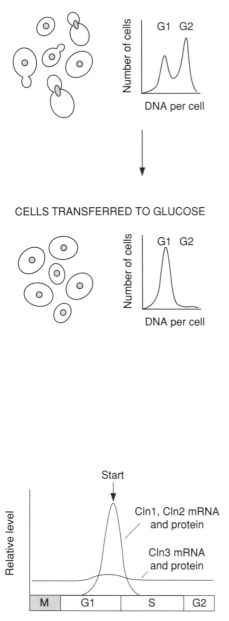

CELLS TRANSFERRED TO GLUCOSE

FIGURE 6-4   Regulation of G1 cyclins. The *CLN1* and *CLN2* genes are transcribed only during G1, so their protein products are present only in this part of the cell cycle. The *CLN3* gene is transcribed throughout the cell cycle, so its product is always present. The activity of the Cln3 protein is regulated by post-translational modification. Reference: Wittenberg, C., Sugimoto, K., & Reed, S. I. *Cell* **62**, 225–237 (1990).

FIGURE 6-3   G1 cyclins induce Start. Budding yeast strains that can express G1 cyclins only under certain conditions show that these proteins are required for passage through Start. To create such a strain, the normal G1 cyclin genes were all deleted and replaced with a copy of the *CLN2* gene whose transcription is controlled by a galactose-inducible promoter. This promoter is active when cells are grown in the presence of galactose but inhibited by glucose. When cells are switched from galactose to glucose, they arrest at the next Start, demonstrating two points: G1 cyclins are essential for passage through Start, and they are unstable and have to be freshly synthesized in each cell cycle. Reference: Richardson, H. E., Wittenberg, C. W., Cross, F., & Reed, S. I. *Cell* **59**, 1127–1133 (1989).

contained a single G1 cyclin whose expression could be experimentally controlled. To test the role of *CLN2* expression, the *CLN1, CLN2,* and *CLN3* genes were deleted from the chromosome in a strain where the coding sequences of the *CLN2* gene had been placed under the control of an inducible promoter. This promoter is active when cells are grown on galactose-containing medium but inactive when they are grown on glucose. On galactose the strain expresses Cln2 and grows normally, but when it is switched to glucose, cells complete the cell cycle and arrest at the next Start (Figure 6-3). This rapid arrest shows that the G1 cyclins, like their mitotic counterparts, are unstable proteins that must be synthesized anew in each cell cycle.

The behavior of Cln mRNAs and proteins during the cell cycle supports the idea that accumulation of the G1 cyclins regulates Start. Cln1 and Cln2 mRNAs and proteins increase in abundance after the end of mitosis, reaching maximum levels just before Start, and then declining rapidly (Figure 6-4). The abundance of Cln3 mRNA and protein, however, shows only mild fluctuation during the cell cycle. All three G1 cyclins are unstable proteins with a half-life of less than 10 minutes.

We are just beginning to understand the biochemical mechanisms by which the interactions of Cdc28 with G1 cyclins induce cells to pass Start. Protein kinase activity can be detected in Cdc28–G1 cyclin complexes isolated from budding yeast in G1. There is evidence that post-translational modification regulates the activity of Cln3, but we do not know if post-translational modification of Cln1, Cln2, or Cdc28 plays a role in controlling Start.

G1 cyclin transcription is activated by a positive feedback loop

Like mitosis, Start is induced by a positive feedback loop. In mitosis small amounts of MPF stimulate the post-translational reactions that activate preMPF, whereas Cdc28–G1 cyclin complexes induce Start by a pathway that involves both transcriptional and post-translational regulation (Figure 6-5). The synthesis of Cln1 and Cln2 mRNA is stimulated by a transcription factor composed of two proteins, Swi4 and Swi6, whose activity depends on phosphorylation. Active Cdc28–G1 cyclin complexes appear to activate

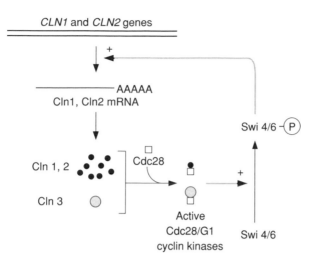

CLN1 and CLN2 genes

Cln1, Cln2 mRNA AAAAA

Cln 1, 2

Cln 3

Cdc28

Active
Cdc28/G1
cyclin kinases

Swi 4/6

Swi 4/6 (P)

FIGURE 6-5 Positive feedback loop for G1 cyclin expression in budding yeast. The transcription of Cln1 and Cln2 is stimulated by a transcription factor made up of the Swi4 and Swi6 proteins, but this factor is fully active only when it is phosphorylated. The phosphorylation of Swi4/6 is induced by complexes of G1 cyclins and Cdc28; thus the expression of a small amount of G1 cyclins can trigger a rapid increase in the synthesis of these proteins. References: Nasmyth, K., & Dirick, L. *Cell* **66**, 995–1013 (1991); Ogas, J., Andrews, B. J., & Herskowitz, I. *Cell* **66**, 1015–1026 (1991); Cross, F. R., & Tinkelenberg, A. H. *Cell* **65**, 875–883 (1991).

CLN1 and CLN2 transcription by inducing the phosphorylation of Swi4/6. Therefore, once G1 cyclins exceed a certain level, they activate some of the Cdc28 in the cell, leading to increased G1 cyclin production, which activates more Cdc28 and leads to rapid and irreversible passage through Start.

Although we understand the basic mechanisms that activate Cdc28 at Start, many important questions remain unanswered. What stimulates the initial low-level transcription of the G1 cyclins that activates the positive feedback loop? Are there proteins that restrain the activation of Cdc28 at Start, much as Wee1 restrains the activation of MPF? When budding yeast cells attain a certain size the level of G1 cyclins increases dramatically. How does achieving this threshold size induce the accumulation of G1 cyclins? How do cells know how big they are? Are they measuring mass, volume, length, or some biochemical parameter that varies in a predictable way with one of the physical parameters? Although a variety of models have been proposed to explain how cells can measure their own size, little experimental evidence exists to support any one of them, and these remain vital unanswered questions in biology.

The activation of G proteins by mating factor receptors leads to cell cycle arrest

Studies of mating-factor-resistant mutants have identified many components of an intracellular signaling pathway. The pathway contains a surprisingly large number of components, whose interactions are complex. We shall see that the pathway by which mating factors arrest the budding yeast cell cycle is strikingly similar to the pathway by which growth factors induce mammalian cells to pass through the restriction point.

The mating factor receptors are proteins that are embedded in the plasma membrane. Receptor molecules with mating factor bound to them can interact with a **G protein** composed of three subunits: α, β and γ. When the protein is in its resting state, the three subunits form a trimeric complex with a

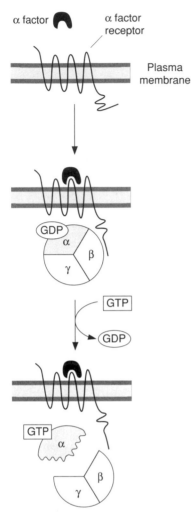

α factor

α factor receptor

Plasma membrane

GDP

α

β

γ

GTP

GDP

GTP

α

β

γ

Free β–γ complex activates signaling pathway

FIGURE 6-6   Mating factors activate a trimeric G protein. Mating factors bind to mating factor receptors that are embedded in the plasma membrane. The mating factor–receptor complexes interact with a trimeric G protein to stimulate the α subunit to exchange its bound GDP for GTP and dissociate from the β and γ subunits. The β–γ complex initiates the signaling pathway that leads to cell cycle arrest and expression of conjugation genes. Reference: Whiteway, M., et al. *Cell* **56,** 467–477 (1989).

molecule of GDP tightly bound to the α subunit. When mating factors bind to their receptors, the resulting complex interacts with the G protein and induces the α subunit to exchange GDP for GTP and dissociate from the β and γ subunits. These two subunits remain bound together and initiate the signaling pathway, probably by binding to and activating the protein kinase encoded by the *STE20*[+] gene (Figure 6-6).

Two features of the G protein play critical roles in the pathway. First, a single mating-factor-receptor complex can induce many G protein molecules to exchange GDP for GTP, thus amplifying the initial extracellular signal. Second, because the free α subunit has a GTPase activity that can hydrolyze its bound GTP to GDP, the intracellular signal disappears rapidly when mating factor is removed. Once the α subunit is in the GDP-bound form, it rebinds to the β–γ complex, terminating the latter's kinase-stimulating activity until the G protein interacts once more with a mating-factor-receptor complex.

The G protein in the mating factor pathway is an example of a trimeric G protein. In the GDP-bound form, the α subunit has a high affinity for the β–γ complex, and the protein exists as an inactive trimer. Upon GTP binding, the α subunit is released. The ability of the free β–γ complex to activate the mating factor signaling pathway is unusual. In most trimeric G proteins, it is the free α subunit, rather than the β–γ pair, that induces the activation of other cellular enzymes. G proteins are members of a very large class of proteins that use the free energy of GTP hydrolysis to regulate their behavior in cells. All of these proteins have weak intrinsic GTPase activity that hydrolyzes bound GTP to GDP, producing a conformational change in the protein that alters its activity. In many cases the GDP remains tightly bound to the protein until it interacts with a **guanine nucleotide exchange protein** that accelerates the exchange of the bound nucleotide with free GTP or GDP. Since cells contain much higher concentrations of GTP than of GDP, the net effect of the exchange proteins is to replace bound GDP with GTP. In the mating pathway, mating-factor-receptor complexes act as exchange proteins, but unoccupied receptors do not.

Tubulin is another example of a protein whose behavior is regulated by GTP hydrolysis. Tubulin dimers can polymerize only when the β subunit carries GTP, but once in microtubules, the GTP is hydrolyzed to GDP. This hydrolysis reaction is crucial for the dynamic instability of microtubules: Only those that contain GDP tubulin can undergo the transition from the growing to the shrinking phase. Perhaps the stability of microtubules is regulated by proteins that regulate the GTPase activity of the tubulin in the polymer.

A G protein activates a protein kinase cascade that induces the destruction of G1 cyclins

The $\beta-\gamma$ complex derived from the receptor-coupled G protein arrests the cell cycle by activating a protein kinase cascade (Figure 6-7). In this pathway each kinase phosphorylates and activates the next kinase in the chain. Eventually the pathway branches, with one fork phosphorylating and activating a transcription factor that induces the transcription of mating-specific genes. The other branch of the pathway activates proteins that destroy or inactivate the G1 cyclins, leading to cell cycle arrest.

The steps immediately after the formation of the $\beta-\gamma$ complex are understood the least. The complex is believed to bind to and activate Ste20 (sterile), a protein kinase that in turn leads to the activation of a protein called Ste5 whose biochemical function is unknown. Ste5 then induces the activation of Ste11, a protein kinase. The steps downstream of Ste11 are more firmly established. Activated Ste11 phosphorylates and activates Ste7, another protein kinase. Activated Ste7 then phosphorylates and activates two other protein kinases, Fus3 and Kss1, which both show homology to the MAP family of protein kinases. Either Fus3 (fusion) or Kss1 (kinase suppressor of super sterile) can phosphorylate and activate Ste12, a protein that induces the transcription of genes involved in conjugation.

Fus3 plays at least two roles in inducing cell cycle arrest. First, it phosphorylates Far1 (factor arrest), which, in its phosphorylated form, binds to the Cln2–Cdc28 and Cln1–Cdc28 complexes and induces the degradation of Cln1 and Cln2 by an unknown mechanism. Second, it may phosphorylate Cln3, inactivating this G1 cyclin. Mating factor probably induces the disappearance of Cln1 both by inhibiting its transcription and by inducing the Far1-dependent degradation of the protein.

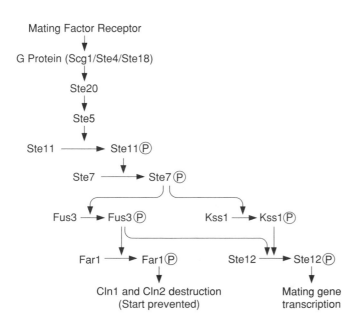

FIGURE 6-7 Mating-factor-induced signaling pathway. Binding of mating factors to their receptors induces a protein kinase cascade that leads to cell cycle arrest and transcription of conjugation genes. The steps between the activation of the G protein and the activation of Ste11 are not well understood. Ste11 phosphorylates Ste7, which in turn phosphorylates Fus3 and Kss1. Active Fus3 or Kss1 can turn on Ste12, a transcription factor that induces expression of conjugation specific genes and boosts the expression of some of the genes in the signaling pathway. Active Fus3 phosphorylates Far1, which induces the degradation of Cln1 Cln2. Fus3 may also induce the phosphorylation and inactivation of Cln3. Reference: Marsh, L., Neiman, A. R., & Herskowitz, I. *Ann. Rev. Cell Biol.* **7**, 699–728 (1991).

The strength of the mating-factor-response pathway varies

The mating-factor-response pathway is not a simple linear progression because many steps in the pathway influence steps above, as well as below, them. Thus, the Ste12 protein that activates the transcription of the mating genes also activates the transcription of *FAR1* and genes that encode components of the signaling pathway. In cells that have not been exposed to mating factor, the level of many components in the response pathway is low, and the initial response to stimulation with mating factor is weak. Exposure of cells to mating factor induces components of the response pathway, leading to an increased response. Regulating the pathway in this way may have two advantages. First, it is economical, since cells do not produce high levels of components in the pathway until they are stimulated by mating factors. Second, the weak initial response may help to ensure that a transient exposure to mating factors does not lead immediately to a lengthy cell cycle arrest. In other words, courtship is gradual, so a cell can recover rapidly if its mating partner disappears.

The mating-factor-response pathway also shows **adaptation:** After the initial increase in responsiveness, continued exposure to mating factor leads to a gradual decrease in the response. This decrease reflects a reduction in the number of mating factor receptors on the surface of the cell and possibly also in the sensitivity of intracellular components of the pathway. We can view adaptation as a mechanism that allows cells that are near cells of the opposite mating type but are unable to mate with them to recover and continue dividing. We expect that the features of an initial increase in responsiveness followed by adaptation are common in response pathways that regulate the cell cycle in response to signals generated by other cells.

Nutrients control the activity of adenylate cyclase

The budding yeast *cdc25^ts* and *cdc35^ts* mutants cease growing and cannot pass through Start at the restrictive temperature (see page 40). Because these mutants behave as if they were starved, even in the presence of adequate supplies of nutrients, they have helped elucidate how cells respond to starvation. *CDC35+* encodes adenylate cyclase, the enzyme that produces **cyclic AMP (cAMP)** from ATP.

The phenotype of *cdc35^ts* mutants demonstrates that cell growth and Start both depend on cAMP. Cyclic AMP regulates the cell cycle by activating the cAMP-dependent protein kinase, which is a tetramer composed of two regulatory and two catalytic subunits (Figure 6-8). In the absence of cAMP, the regulatory subunits bind tightly to the catalytic subunits, inhibiting protein kinase activity. The binding of cAMP to the regulatory subunits induces their dissociation from the catalytic subunits, which can then phosphorylate their substrates. Mutations that inactivate the regulatory subunit make the kinase activity constitutive, even in the absence of cAMP, and allow cells to grow without adenylate cyclase. These mutants cannot arrest the cell cycle prop-

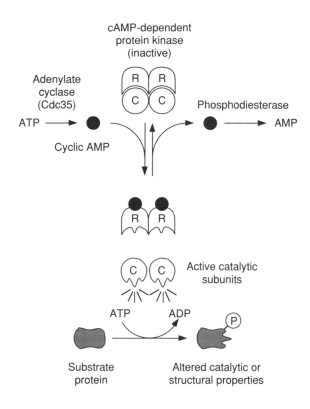

cAMP-dependent
protein kinase
(inactive)

Adenylate
cyclase
(Cdc35)

ATP

Cyclic AMP

Phosphodiesterase

AMP

Active catalytic
subunits

ATP   ADP

Substrate
protein

Altered catalytic or
structural properties

FIGURE 6-8 Regulation of cyclic AMP-dependent protein kinase. Adenylate cyclase (the product of the *CDC35⁺* gene) catalyzes the formation of cyclic AMP from ATP. Cyclic AMP binds to the regulatory subunits of cyclic AMP-dependent protein kinase, inducing them to dissociate from the catalytic subunits. The free catalytic subunits phosphorylate their substrates, producing changes in their structural or catalytic properties or both. Cyclic AMP is destroyed by phosphodiesterases (encoded by the *PDE1⁺* or *PDE2⁺* genes).

erly when they are starved, demonstrating that in budding yeast, nutrient availability regulates the activity of cAMP-dependent protein kinase.

Reducing the activity of cAMP-dependent protein kinase decreases the rate of protein synthesis by an unknown mechanism. Mutations that increase the half-life of G1 cyclins impair the ability of starvation to arrest the cell cycle, suggesting that it is the rate of cyclin synthesis that determines whether cells can pass Start.

Unstable proteins, like the G1 cyclins, are especially well suited to regulate biological processes. The concentration of a protein in a cell is determined by its rates of synthesis and degradation. Any protein will eventually reach a steady-state concentration at which the rate of synthesis exactly balances that of degradation. When either rate changes, the speed with which the protein attains a new steady-state concentration is determined by the rate of degradation. The more rapidly a protein is degraded, the more rapidly it reaches the new steady state concentration (Figure 6-9). Thus, for a protein with a short half-life, a twofold reduction in synthesis rate will lead rapidly to a twofold reduction in the steady state concentration.

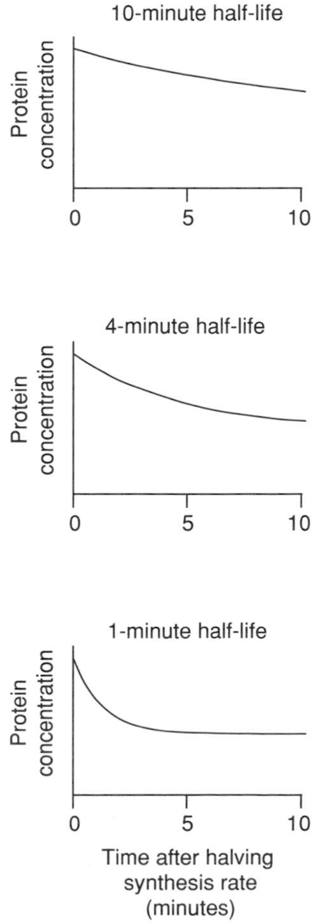

FIGURE 6-9 Regulation of unstable proteins. The half-life of a protein determines how fast its concentration changes when the rate of synthesis is halved. The shorter the half-life, the more rapidly the concentration of the protein responds to a reduction in its synthesis.

10-minute half-life

Protein
concentration

0   5   10

4-minute half-life

Protein
concentration

0   5   10

1-minute half-life

Protein
concentration

0   5   10

Time after halving
synthesis rate
(minutes)

Ras, a small G protein, regulates adenylate cyclase in budding yeast

Studies on genes that cause cancer have shown that a **small G protein** called **Ras** (rat sarcoma virus) regulates cell proliferation (see pages 156–157). Small G proteins have a single subunit of about 21 kd and are involved in many different aspects of cell biology, including cellular signaling pathways, the traffic of proteins between different membrane-bounded compartments of the cell, and protein synthesis. Like the trimeric G proteins, small G proteins are active in their GTP-bound form.

Isolating homologs of Ras in yeast has emphasized that although many important regulators of the cell cycle are strongly conserved between different cell types, their primary functions are not always the same. Human Ras can substitute for yeast Ras, proteins, but the primary action of Ras differs in budding yeast cells and vertebrate cells. In yeast Ras regulates adenylate cyclase (Figure 6-10), but in mammalian cells it has some other unknown target (see pages 109–110).

Budding yeast has two Ras proteins encoded by the $RAS1^+$ and $RAS2^+$ genes, and strains that lack both genes are inviable. Dominant RAS mutants that decrease the intrinsic GTPase activity of Ras have high levels of cAMP and fail to arrest in G0 when they are starved. The budding yeast $CDC25^+$ gene encodes the guanine nucleotide exchange protein that activates Ras by stimulating the displacement of GDP and the binding of GTP. Mutations in this gene arrest the cell cycle at Start because Ras accumulates in its inactive, GDP-bound form.

Compared with the trimeric G proteins, small G proteins have very weak intrinsic GTPase activity, which can be greatly increased by **GTPase activating proteins (GAPs)** that interact with small G proteins. The budding yeast $IRA1^+$ and $IRA2^+$ genes encode GTPase-activating proteins that act on Ras.

FIGURE 6-10 Ras controls adenylate cyclase. In budding yeast GTP-bound Ras activates adenylate cyclase to stimulate the formation of cAMP. Ras is activated by the budding yeast Cdc25 protein—which is a nucleotide exchange protein (not to be confused with the fission yeast Cdc25, which is a tyrosine phosphatase)—and inactivated by Ira1 or Ira2, a protein that stimulates the GTPase activity of Ras. Reference: Broach, J. R. *Curr. Opin. Genet. Dev.* **1,** 370–377 (1991).

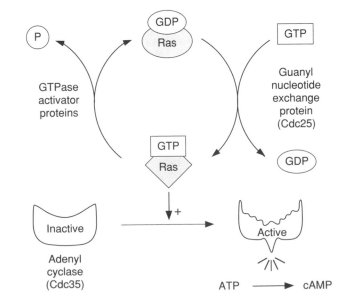

Mutations that inactivate both *IRA* genes behave like dominant *RAS* mutations: They slow hydrolysis of the GTP bound to Ras, increasing cAMP levels and abolishing the ability of starvation to arrest the cell cycle in G0. Thus, controlling the balance between nucleotide exchange proteins and GTPase-activating proteins regulates the activity of small G proteins. Increasing the activity of nucleotide exchange proteins stimulates G protein activity, whereas increasing the activity of GTPase activating proteins inhibits G protein activity.

Although we understand the outlines of how nutrition regulates the budding yeast cell cycle, important questions remain about the details of this control. How do nutrient levels control the activity of the *CDC25, IRA1,* and *IRA2* gene products, thereby controlling Ras activity and cAMP levels? Which substrates of cAMP-dependent protein kinases alter the rate of protein synthesis and prevent G1 cyclins from accumulating to the levels required to induce Start? Finally, genetic evidence suggests that Ras has other targets besides adenylate cyclase. What are these targets, and do they correspond to the primary targets of Ras in mammalian cells?

Reduced cAMP levels induce meiosis

Diploid budding yeast cells enter the meiotic cell cycle and form spores when they are simultaneously starved of nitrogen and given carbon sources, such as acetate and ethanol, that have to be oxidized rather than fermented (see pages 34–36). Shifting diploid *cdc25^{ts}* or *cdc35^{ts}* mutants growing in rich medium to the restrictive temperature decreases cAMP levels and makes cells enter the meiotic cell cycle, demonstrating that sporulation, like G0

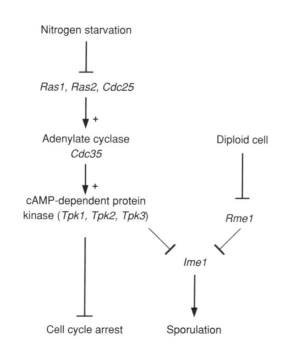

FIGURE 6-11 Control of sporulation in budding yeast. The key gene in the induction of sporulation is *IME1⁺*. *IME1⁺* expression is inhibited by Rme1, a protein that is expressed only in haploids, and by cAMP-dependent protein kinases. Nitrogen starvation leads to the inactivation of cAMP-dependent protein kinase, allowing *IME1⁺* to be expressed in diploid cells. The induction of *IME1⁺* expression leads to the expression of sporulation-specific genes, which can induce cells that have not yet begun DNA replication to enter the meiotic cell cycle. Reference: Malone, R. E. *Cell* **61**, 375–378 (1990).

arrest, is induced by reducing the level of cAMP. By contrast, mutations that lack the regulatory subunit of the cAMP-dependent protein kinase and have constitutively high kinase activity cannot enter the meiotic cycle under any conditions. In budding yeast the cAMP-dependent protein kinase blocks sporulation in part by inhibiting the expression of *IME1+* (inducer of meiosis), which normally induces sporulation. The regulation of *IME+* expression also explains why haploid cells cannot sporulate. Haploids express *RME1+* (repressor of meiosis), an inhibitor of *IME+* transcription. In diploids the products of the *MATa* and *MATα* genes combine to form a transcriptional repressor that inhibits transcription of *RME1+*, thus removing one of the blocks to the expression of *IME1+* and entry into the meiotic cell cycle (Figure 6-11).

In the fission yeast sporulation occurs immediately after mating and requires the mating factor signaling pathway as well as a reduction in cyclic AMP levels. In this organism the activity of adenyl cyclase is not directly regulated by Ras.

GROWTH CONTROL IN MAMMALIAN CELLS

In budding yeast, simple nutrients control the level of G1 cyclins in the cell and regulate passage through Start. Mammalian cells are continuously supplied with nutrients, and cell growth and proliferation are regulated by growth factors, which act as specific signaling molecules. Are the signaling pathways in mammalian cells similar to those in yeast, and to what extent do mammalian growth factors regulate passage through the restriction point by controlling the accumulation of G1 cyclins?

The cell cycle in mammalian cells varies greatly

In complex multicellular organisms different cells divide at very different rates. For example, the cells that line our intestines live only three days and must be constantly replaced by the division of precursor cells. On the other hand, the life span of liver cells is more than a year, so cell division in this organ is rare.

The variability in the length of the cell cycle occurs mainly in G1 and G2 and reflects the ability of cells to exit from the cell cycle during either of these intervals (Figure 6-12). Cells in G0 have left the cycle after division but before

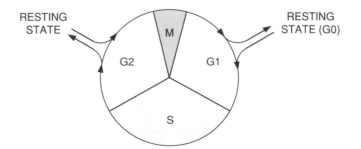

FIGURE 6-12 Cell cycle exit in G1 and G2. Many cells can withdraw from the cell cycle, entering G0 or a stable G2 arrest, in which growth is also halted.

the restriction point and account for most of the nongrowing, nonproliferating cells in the human body. Some cells (such as some epidermal cells) leave the cycle during G2 and arrest without growth or proliferation. There is no specific name for this arrest phase, but in many respects the G0 and G2 arrests are similar, and their underlying mechanism is likely to be the same. Because it is difficult to assess the cell cycle state of cells in intact organs, there is considerable controversy over what fraction of cells in adult mammals are arrested in G0 and what fraction are arrested in G2. We do know that neurons leave the cell cycle permanently and never divide again, although we do not understand the mechanism of this irreversible exit from the cell cycle.

Mammalian cell proliferation is studied in tissue culture

It is difficult to study cell proliferation in intact multicellular animals; therefore, most studies are performed on cells proliferating in culture. Animal cells were first cultured in 1907 by placing a piece of tissue (called an explant) in a glass dish filled with **serum,** the fluid produced when blood clots. In such cultures some cells migrate away from the tissue along the surface of the dish, and proliferate (Figure 6-13). These cells are usually **fibroblasts,** which play important roles in the formation and maintenance of connective tissue. The cells growing out of explants are called **primary cells,** and have properties that closely resemble those of cells in vivo. They will proliferate only if they are firmly attached to and spread out over an adhesive substrate, a property called **anchorage dependence.**

Primary cells continue to proliferate until the dish is covered by a single layer of cells, but once the cells form a continuous sheet they stop dividing. This phenomenon is called **contact inhibition** of cell proliferation, and its basis is poorly understood. Three mechanisms for contact inhibition have

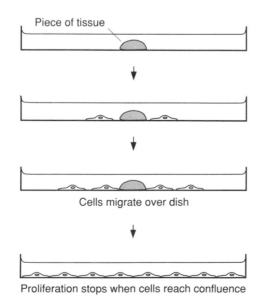

FIGURE 6-13 Tissue culture. The proliferation of vertebrate cells is usually studied in tissue culture. Such cultures are initiated by placing a piece of tissue in a glass dish that contains serum. Cells migrate out of the tissue along the surface of the glass and proliferate. Proliferation ceases when the dish is covered with a single layer of cells.

been proposed: competition between neighboring cells for the growth factors that are present in trace amounts in the medium, the binding of molecules on the surface of one cell to receptors on an adjacent cell to activate signaling pathways that inhibit cell proliferation, and the ability of cells to secrete molecules that inhibit their own proliferation. Without understanding the mechanism of contact inhibition in vitro, it is impossible to assess the relevance of this phenomenon to the control of cell proliferation in vivo.

Gentle protease treatment removes the contact inhibited primary cells from the dish, and transferring them at lower densities to new dishes that contain fresh medium induces them to resume proliferation. This process can be continued until the cells have divided about 50 times. At this point the culture starts to **senesce:** fewer and fewer cells divide, and cells begin to die (Figure 6-14). A few of the cells in the culture accumulate genetic changes that allow them to avoid this fate. These cells can proliferate indefinitely, and homogeneous populations called **tissue culture lines** can be established from single isolated cells. As long as they are transferred to new dishes frequently and not allowed to grow to high densities, these tissue culture lines remain anchorage-dependent and subject to contact inhibition, continue to resemble primary cells in their growth control, and are said to be nontransformed. But though established cell lines superficially resemble primary cells and cells in

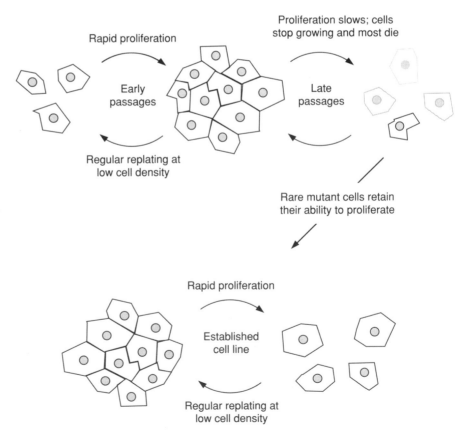

FIGURE 6-14 Senescence. Cells isolated from normal animals cannot proliferate forever. For cells isolated from young mammals, proliferation slows and cells begin to die after about 50 divisions in tissue culture. This phenomenon, called senescence, occurs after a smaller number of cell divisions in cultures isolated from older individuals. Eventually some of the cells in the culture accumulate genetic changes that allow them to escape senescence. As long as they are transferred to new dishes periodically, these cells can divide indefinitely, and tissue culture cell lines can be established from the progeny of a single cell. Reference: Hayflick, R. *Exp. Cell Res.* **37**, 614–636 (1965).

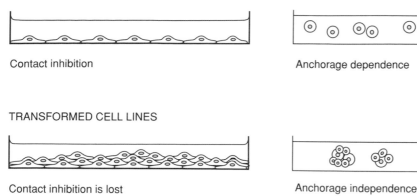

NORMAL CELL LINES

Contact inhibition

Anchorage dependence

TRANSFORMED CELL LINES

Contact inhibition is lost

Anchorage independence

FIGURE 6-15 Transformed cell lines. Primary cells and normal tissue culture cell lines show two forms of growth restriction: Cells cannot proliferate if they are touching other cells on all sides (contact inhibition) or if they are not spread out on a solid substrate (anchorage dependence). Transformed cell lines lack either of these growth controls: They can grow on top of each other and in suspension.

vivo, they carry mutations that have altered the details of proliferation control enough to allow them to escape senescence. Some of these changes may involve the activation of genes that inhibit the induction of cell death by inappropriate growth stimuli (see pages 144–145).

When tissue culture cell lines are allowed to grow repeatedly to high densities, **transformed cell lines** that have lost normal growth controls are selected. The proliferation of these lines is anchorage-independent and is not contact inhibited (Figure 6-15). When injected into animals without functional immune systems, transformed cell lines cause tumors. Transformed cell lines can also be produced by allowing cells to grow out of tumors or by treating primary cells or nontransformed cell lines with chemical carcinogens or cancer-inducing viruses.

Studying the differences between transformed and nontransformed cell lines has revealed two classes of genes involved in regulating cell proliferation (see Chapter 9). One class, called **proto-oncogenes,** can suffer dominant mutations that convert them into **oncogenes** that transform cells. Many proto-oncogenes encode components of the signaling pathway that converts binding of growth factors at the cell surface into stimulation of the cell cycle engine. The dominant mutations that activate these genes allow cells to escape from the normal controls of proliferation. Members of the second class of genes are called **tumor suppressor genes** because their products act as cell brakes that tend to inhibit cell proliferation. Recessive mutations that inactivate tumor supressor genes can transform cells by leading to unrestrained proliferation, and inactivating controls that help cells avoid genetic damage (see Chapter 9).

Serum supplies tissue culture cells with growth factors

For many years tissue culture cells had to be grown in medium that contained serum. Even when all known low-molecular-weight nutrients and vitamins were included in the medium, cells would arrest in G0 unless the medium also contained some serum. Restoring serum to the medium allowed cells to grow and proliferate again, showing that serum contains growth factors essential for these processes. Starving cells of a particular amino acid or partially inhibiting protein synthesis had the same effect as serum deprivation: All three treatments arrested cells in G0. Together these observations led to the definition of the restriction point as a checkpoint in G1, analogous to Start in yeasts, at which the cell cycle was arrested under suboptimal conditions.

During the last 15 years biochemists have identified and purified the growth factors in serum. The existence of **platelet-derived growth factor (PDGF)** was deduced from the observation that plasma, the cell-free fluid derived from unclotted blood, does not induce cell proliferation, whereas serum induces proliferation of many tissue culture cell lines. The ability of serum to support cell proliferation suggested that the process of clotting liberates some growth factor. Since platelets liberate many of their components during clotting, it seemed likely that they would contain a growth factor. The contents of purified platelets stimulated cell proliferation, confirming this hypothesis. Starting with 200 liters of human blood, investigators purified PDGF to homogeneity and showed that it is a small family of dimeric proteins composed of two identical or closely related subunits, one of which is the product of the *c-sis* (simian sarcoma) proto-oncogene. The release of PDGF during blood clotting probably plays an important part in inducing the cell proliferation that occurs during wound healing.

The struggle to purify growth factors reflects their ability to control cell proliferation at vanishingly low concentrations — in the range from 10^{-9} to 10^{-11} M. Many protein growth factors that control cell growth and proliferation have been identified, and many different cell types can now be induced to proliferate by cocktails of purified growth factors in place of serum. In general, transformed cell lines show sharply reduced requirements for growth factors. There are two classes of growth factors: Broad-range factors, like PDGF, affect the proliferation of many cell types; narrow-range factors affect the proliferation of a single cell type. For example, erythropoietin specifically stimulates the proliferation of red blood cell precursors.

Most growth factors stimulate the growth and proliferation of cells that can detect them. Starving cells of such growth factors induces a rapid decrease in the rate of protein synthesis and prevents cells that have not already passed the restriction point from doing so. A few growth factors have different effects on different cell types: For example, transforming growth factor β (TGF-β) stimulates some cell types to proliferate, but in other cell types it inhibits the ability of other growth factors to induce proliferation. There are no mammalian examples of growth factors that, like the yeast mating factors, act solely to inhibit cell proliferation.

Many vertebrate growth factors stimulate cell differentiation in particular cell types. The ability of vertebrate growth factors to control both cell proliferation and cell differentiation, parallels the ability of yeast mating factors to induce both cell cycle arrest and the expression of genes involved in the physical process of mating.

The restriction point marks the end of a process that begins with the birth of a cell

Following the fate of individual cells in a population by time lapse cinematography permitted a detailed analysis of passage through the restriction point. In a typical experiment, an asynchronous population of proliferating cells was filmed for 24 hours before a 1-hour period of serum deprivation and then for 36 hours afterward. Two parameters were measured for each cell: the time between the birth of the cell and the onset of serum deprivation and the duration of the cell cycle in which the period of serum starvation occurred. Comparing the responses of many different cells in the population revealed the effects of starving cells at different points in the cell cycle (Figure 6-16).

Cells that were more than 4 hours old at the time of serum starvation divided at the same time as they would have without starvation. Cells that were less than 4 hours old at the time of starvation, however, showed a marked delay before their next division. Surprisingly, the delay was much longer than the period of starvation, and the length of this extra delay was independent of the duration of starvation. These painstaking experiments led to two conclusions. First, they showed that the restriction point occurs 4 hours after mitosis in this cell line. Second, the extra delay after serum was restored suggested that cells starved before the restriction point leave the cell

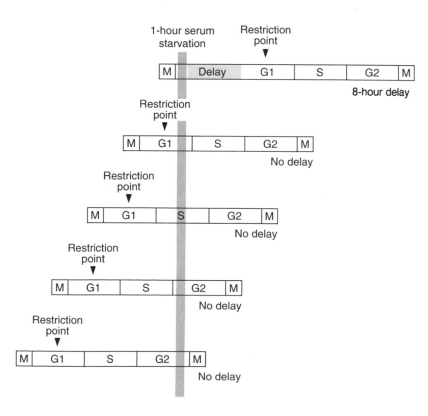

FIGURE 6-16 Identifying the restriction point. The existence of a restriction point for cell proliferation was clearly revealed by analyzing the behavior of individual cells in a population that was filmed for more than one cell cycle. An asynchronous culture of mammalian tissue culture cells was subjected to a 1-hour period of serum deprivation. Because the cells were filmed before and after starvation, the age of the cells at the time of starvation was known, and the effect of starvation on the cell cycle could be measured. Cells starved before the restriction point enter G0 and were delayed in the cell cycle during which starvation occurred. Cells starved after the restriction point showed no delay in completing the cell cycle. Reference: Zetterburg, A., & Larson, O. *Proc. Natl. Acd. Sci. USA* **82**, 5365–5369 (1985).

cycle and enter G0. The additional delay probably represents the time it takes to enter G0 and then return to G1.

THE MOLECULAR BASIS OF GROWTH AND PROLIFERATION CONTROL IN MAMMALIAN CELLS

Having introduced the experimental strategies for studying growth and proliferation, we must now try to synthesize a picture of how mammalian cells control these processes. Although we know much about some of the steps involved in the regulation of proliferation, our ignorance about others keeps us from fitting the steps together into a coherent picture of how cell multiplication is regulated in tissue culture, let alone in intact organisms. As a result much guesswork is involved in the description that follows.

We do not understand the details of how cells control their growth. In growing cells, protein synthesis exceeds protein degradation, and the cell increases in mass. When cells are not growing, protein synthesis and degradation are precisely balanced, and the cell mass remains constant. We know that removing growth factors reduces the rate of protein synthesis, but does it also increase the rate of protein degradation? At a molecular level, how are the rates of protein synthesis and degradation controlled? How does the pattern of protein synthesis change as the rate of protein synthesis changes? How does the pattern of protein degradation change as the rate of protein degradation changes? Are rates of ribosomal RNA, tRNA, and lipid synthesis regulated as a consequence of changes in the rate of protein synthesis, or are they controlled independently? Surprisingly, we know the answer to very few of these questions. All we know for certain is that when cells are deprived of growth factors, they reduce their rate of protein synthesis, stop growing, and lose the ability to pass the restriction point.

Cell proliferation is regulated by a chain of events that begins with the interaction of growth factors with specific **growth factor receptors** located in the plasma membrane. The binding of the growth factors induces the receptors to generate signals inside the cell, including **second messengers,** small molecules that activate a variety of intracellular enzymes, mainly protein kinases and phosphatases. Changes in protein phosphorylation then lead to the transcription of **early response genes** that encode transcription factors, which in turn induce the transcription of the **delayed response genes.** The products of the delayed response genes include G1 cyclins and relatives of Cdc2/28 that interact with each other to yield active protein kinases. These kinases are believed to perform two functions: They inactivate the tumor suppressor gene products that restrain the cell cycle engine, and they activate the DNA replication machinery, thereby inducing passage through the restriction point. Rather than attempting a general survey of how cell proliferation is regulated in different cell types, we have chosen to focus on the details of the response to PDGF.

FIGURE 6-17 Activation of the platelet-derived growth factor (PDGF) receptor. Binding of a dimeric PDGF molecule brings two PDGF receptors close together so that they can phosphorylate and activate each other. The tyrosine phosphates on the receptor are mooring sites at which other proteins that regulate cell proliferation bind and become active. These proteins include phosphatidylinositol-3-kinase (PI$_3$K), phospholipase C–γ (PLC–γ), and the Ras GTPase-activating protein (not shown). Reference: Williams, L. T., Escobedo, J. A., Fantl, W. J., Turck, C. W., & Klippel, A. *Cold Spring Harbor Symp. Quant. Biol.* **56**, 243–250 (1991).

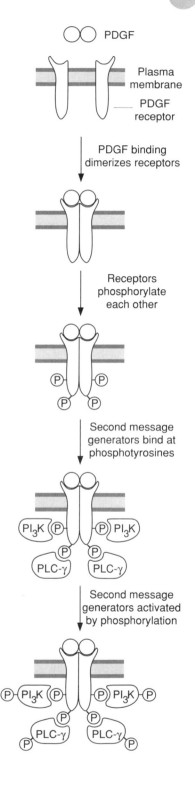

The PDGF receptor is a transmembrane protein kinase

The PDGF receptor has an extracellular domain that binds PDGF and an intracellular domain that is a protein kinase. Binding of PDGF induces the dimerization of the receptor both because PDGF is a dimer that can bind two receptors and because binding of PDGF induces conformational changes in the receptor that favor dimerization. Once the receptor has dimerized the kinase domain of one receptor phosphorylates the neighboring receptor on a number of tyrosine residues.

The autophosphorylation of the PDGF receptor has two consequences; Phosphorylation at some sites activates the protein kinase activity of the PDGF receptor; phosphates at other sites create mooring sites for proteins involved in the generation of second messengers (Figure 6-17). These proteins contain a specialized protein domain, called the **SH2** (src homology 2) domain, that recognizes and binds to phosphorylated tyrosine residues. Binding of target proteins to the PDGF receptor activates the proteins indirectly by making them easy targets for phosphorylation by the receptor's tyrosine kinase. The SH2 domains in different proteins recognize different phosphotyrosine residues in the intracellular domain of the PDGF receptor.

Many other growth factor receptors are also transmembrane tyrosine kinases whose mode of action is broadly similar to that of PDGF. Others, such as receptors for the specialized growth factors that regulate the production of B and T lymphocytes, are not themselves tyrosine kinases but interact with and activate cytoplasmic tyrosine kinases that act in the same way as the kinase domain of the PDGF receptor.

The activated PDGF receptor induces the production of second messengers

The proteins that bind the phosphorylated tyrosines of the PDGF receptor are key regulators of second messenger production. One such protein is **phospholipase C–γ,** which cleaves particular membrane lipids, called **phosphoinositides.** Both of the cleavage products, diacylglycerol and inositol trisphosphate, are potent second messengers (Figure 6-18). Diacylglycerol activates a protein kinase called **protein kinase C,** while inositol trisphosphate stimulates the release of calcium from stores in the endoplasmic reticulum, leading to increased cytoplasmic calcium concentrations. Most of the

FIGURE 6-18 Phospholipase C–γ generates two second messengers. Phospholipase C–γ hydrolyzes phosphoinositides in the plasma membrane
to produce diacylglycerol and inositol
trisphosphate (IP₃), two potent second
messenger molecules. Diacylglycerol
activates protein kinase C; inositol trisphosphate induces the liberation of calcium from membrane-bound stores. The
liberated calcium activates several proteins, including protein kinase C and
the calcium calmodulin-dependent kinase.

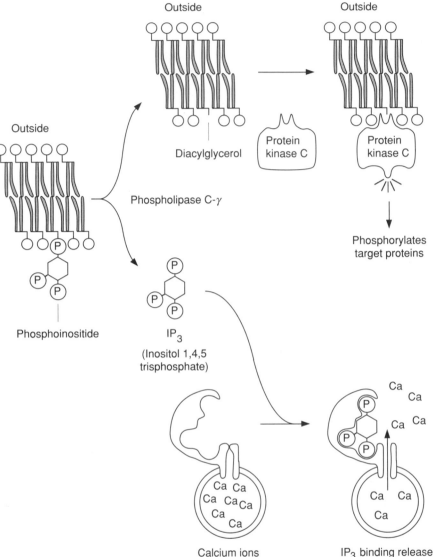

effects of calcium are mediated by a small protein called **calmodulin.** Calmodulin lacks enzymatic activity but binds to and stimulates the activity of a
number of enzymes, including protein kinases, phosphatases, and proteases.
The substrates of protein kinase C and other calcium-stimulated enzymes that
regulate cellular proliferation are unknown.

   At least two molecules, in addition to phospholipase C–γ, bind to the
phosphorylated PDGF receptor. One is the GTPase-activating protein that
activates the weak intrinsic GTPase activity of the mammalian Ras proteins,
thus inactivating them. Another is **phosphatidylinositol-3-kinase,** which
phosphorylates the 3 position of the inositol ring in phosphoinositides. Al-

though this phosphorylation plays a crucial role in controlling cell proliferation, the mechanism it uses to do so is not understood.

Which of the several proteins that interact with the PDGF receptor induce cell proliferation? One approach to this problem exploits the discovery that the SH2 domain in phosphatidylinositol-3-kinase recognizes a phosphotyrosine in the PDGF receptor that is different from the one recognized by the SH2 phosphate-binding domain in the GTPase-activating protein. Mutant PDGF receptors were constructed that replaced individual tyrosines with nonphosphorylatable phenylalanines. The mutant receptors were introduced into cells that lacked an endogenous PDGF receptor, and the resulting cell lines were treated with PDGF. Receptor mutants that could not interact with the GTPase activating protein still stimulated cell proliferation, suggesting that the activation of this protein is not essential for proliferation. By contrast, receptor mutants that cannot interact with phosphatidylinositol-3-kinase fail to stimulate cell proliferation, suggesting that the activation of phosphatidylinositol-3-kinase (or some other protein that binds to the mutated tyrosine) is required to induce proliferation. Different second messengers probably control the proliferation of different cell types, with the result that the ability of a receptor to bind a particular enzyme that generates a second messenger will be essential in some cells and dispensable in others.

The activation of Ras induces cell proliferation

A variety of experiments indicate that in mammalian cells, as in budding yeast, the activation of Ras plays a crucial role in regulating cell proliferation. A variety of experiments, in mammalian and fission yeast cells, however, indicate that Ras does not act by increasing cAMP levels as it does in budding yeast (see pages 98–99). Despite their apparently different function, mammalian Ras genes substitute, albeit imperfectly, for those of yeast. One possible explanation of this discrepancy is that in budding yeast, Ras activates a minor pathway that is unrelated to cAMP and that this activity of Ras is the sole activity that induces proliferation of mammalian cells.

Recent research has provided a tantalizing glimpse of the pathway that activates Ras. Mammalian cells contain a protein called **GRB2** (Figure 6-19) that contains both SH2 and **SH3** (Src homology 3) domains. The SH2 domains bind to phosphotyrosines in the cytoplasmic domain of the epidermal growth factor receptor, and the SH3 domains bind to guanine nucleotide-exchange proteins and GTPase-activating proteins that regulate the activity of small G proteins such as Ras. Perhaps the binding of the SH2 domains of GRB2 to phosphotyrosines induces conformational changes in the SH3 domains that allow them to activate the guanine nucleotide-exchange proteins that convert Ras into its active form. This idea is supported by the results of a genetic analysis of vulval development in the nematode *Caenorhabditis elegans*. This process requires a cascade of interactions in which a homolog of the epidermal growth factor receptor functions upstream of a homolog of GRB2, which in turn functions upstream of a Ras homolog. Although these results are exciting, there may be other undiscovered intermediaries between the phos-

FIGURE 6-19 A protein kinase cascade induces proliferation. Binding of growth factors to the cell surface induces cell proliferation via a signaling pathway that converts signals at the plasma membrane to new gene expression. Receptor binding leads to the binding of adaptor proteins (such as GRB2 and SEM5) to growth factor receptors. The adaptors in turn activate Ras, probably by stimulating guanine nucleotide exchange proteins that are homologous to budding yeast Cdc25 and *Drosophila* Sos (son of sevenless) proteins. Ras in turn induces the activation and phosphorylation of Raf, a protein kinase. The molecular details of the steps that activate Raf are not well understood. Raf is the first member of the protein kinase cascade, which ultimately leads to the phosphorylation and activation of MAP kinase. Activation of MAP kinase leads to its translocation into the nucleus, where it induces transcription. Note the similarity to the later stages of the pathway that induces the expression of mating genes in yeast (see Figure 6-7). Raf is analogous to Ste5, MAP kinase kinase is homologous to Ste11, MAP kinase is homologous to Fus3 and Kss1, and the mammalian transcription factors are analogous to Ste12. The initial stages of the mating pathway (stippled box) are not homologous to those of the growth factor response pathway.

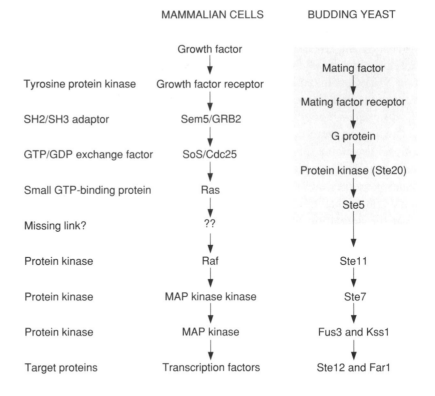

phorylation of growth factor receptors and the activation of Ras, as well as alternative pathways for Ras activation that proceed through entirely different sets of molecules.

Ras activation initiates a protein kinase cascade

The mechanism by which activated Ras induces cell proliferation is even more mysterious than the reactions that activate Ras. We know that activating Ras leads to phosphorylation and activation of a protein kinase called Raf that phosphorylates serines and threonines, but we do not know the nature of the kinase that phosphorylates Raf or how it is activated by Ras.

The activation of Raf in turn initiates a protein kinase cascade that is strikingly similar to the cascade activated by mating factors in budding yeast (see Figure 6-19). The overall organization of Raf and the budding yeast Ste11 protein is similar, although the sequences of the two molecules are not particularly homologous. Activated Raf phosphorylates and activates MAP-kinase kinase (homologous to Ste11), which is a protein kinase that phosphorylates and activates MAP kinase (homologous to Fus3 and Kss1). The activation of MAP kinase has two consequences: translocation of active MAP kinase into the nucleus and the phosphorylation and activation of one member of the S6 kinase family, a group of protein kinases that phosphorylate ribosomal protein S6 and other substrates. We shall see that the presence of active MAP kinase in the nucleus plays an important role in inducing the transcription of genes that induce cell proliferation.

In fission yeast mating factors activate a signaling pathway that shares features with those of budding yeast and mammalian cells. As they do in budding yeast, the receptors for mating type interact with a G protein, but in fission yeast it is the α subunit of the G protein that activates the next step of the pathway. Later in the pathway the activation of a guanine nucleotide-exchange factor activates Ras, which in turn initiates a protein kinase cascade whose members are homologous to those found in budding yeast and mammalian cells. Comparing the different signaling pathways reveals that multi-step pathways have great evolutionary flexibility. Fission yeast and mammalian cells combine homologous components to produce pathways that have a similar organization but very different effects on the cell cycle, whereas budding yeast and fission yeast use a different set of interactions between different components to achieve very similar ends.

Growth factors induce expression of early and delayed response genes

One way of finding genes involved in proliferation control is to look for genes whose expression is induced by growth factors. Comparing the mRNAs in serum-starved cells with those in serum-treated cells identifies two classes of growth-factor-induced genes: early response and delayed response genes (Figure 6-20). The mRNAs of early response genes are undetectable before serum addition, are induced to high levels within 30 minutes of growth factor treatment, and then fall to lower levels that are maintained as long as growth factors are present. In cells treated with protein synthesis inhibitors, the induction of early response genes occurs normally, but the subsequent decline to lower levels of expression is blocked. Many of the early response genes are either known transcription factors or relatives of known transcription factors. Delayed response genes appear more slowly after serum addition, and their transcription depends on protein synthesis, supporting the idea that the early response genes encode proteins that stimulate transcription of the delayed response genes.

Most early and delayed response genes do not show dramatic fluctuations in their mRNA or protein levels as exponentially growing cells pass through the cell cycle. As long as cells are supplied with growth factors, early response genes are expressed and stimulate the transcription of the delayed response genes. Because the mRNA and proteins of the early response genes are unstable, they disappear rapidly when growth factors are removed, halting the transcription of the delayed response genes and inducing entry into G0. When growth factors are restored to the medium, cells cannot progress toward the restriction point until expression of the early response genes restores transcription of the delayed response genes. The time required to induce the full expression of the delayed response genes probably accounts for the slow exit from G0 (see page 105).

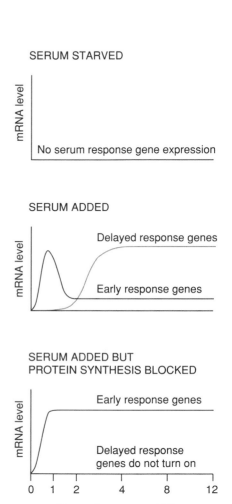

FIGURE 6-20 Early and delayed response genes. Adding serum to serum-starved cells induces the expression of two classes of genes. Early response genes are induced immediately after the addition of serum, and their induction does not require protein synthesis. The transcripts of delayed response genes appear more slowly. Inhibiting protein synthesis with the drug cycloheximide prevents expression of delayed response genes but does not prevent early gene expression.

Active MAP kinase induces transcription of early response genes

In budding yeast, transcription factors induce the expression of G1 cyclins and are themselves activated by Cdc28 – G1 cyclin complexes. Is there a similar autocatalytic loop in the activation of G1 cyclin expression in mammalian cells? We cannot yet answer this question, but the discovery that many of the early response genes are transcription factors suggests that positive feedback loops may play an important role in inducing the restriction point. Three prominent early response genes are the products of the *c-fos, c-jun* and *c-myc* proto-oncogenes. All three of these genes are transcription factors, and Fos and Jun act by forming a dimeric complex that binds to specific DNA sequences and stimulates transcription.

Studies on the regulation of *fos* transcription have revealed how the protein kinase cascade initiated by growth factors can lead to early gene expression. The transcription of *c-fos* appears to be regulated by two proteins, the serum response factor (SRF) and Elk-1, which both bind to the promoter of the *c-fos* gene. Both of these proteins are bound to the promoter in quiescent cells, but the promoter cannot be activated unless the cells are stimulated with serum, which leads to the phosphorylation of Elk-1 (Figure 6-21). In vitro this phosphorylation can be catalyzed by MAP kinase, and mutations in the Elk-1 that block the phosphorylation cannot activate transcription. These results demonstrate for the first time a direct biochemical connection between the cytoplasmic events initiated by growth factors and the transcriptional changes that are believed to induce cell proliferation. An important area for future research is determining how particular cell types respond to growth factors. Possible sources of variability include the presence of different receptors in different cell types, details in the coupling of receptors to second messenger generation and the nature of the signaling pathways that they initiate, and the ability of the same signaling pathway to induce the expression of different early response genes in different cell types.

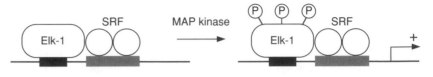

UNSTIMULATED CELL   SERUM-STIMULATED CELL

FIGURE 6-21 Induction of *c-fos. c-fos* is an early response gene that encodes a transcription factor whose expression requires the binding of two proteins, Elk-1 and the serum response factor (SRF), to its promoter. Although these proteins are bound to the promoter in G0 cells, the activation of Elk-1 requires phosphorylation. This modification is catalyzed by MAP kinase, which lies at the end of the growth-factor-induced protein kinase cascade. Reference: Treisman, R., Marais, R., & Wynne, J. *Embo. J.* **11**, 4631 – 4640 (1992).

Delayed response genes include G1 cyclins and relatives of Cdc2/28

In budding yeast, the activated Swi4/6 complex drives cells through Start by inducing transcription of G1 cyclins (see Figure 6-5). Do proteins like Myc play an analogous role in mammalian cells? Recent experiments reveal that G1 cyclins do exist in mammalian cells, but their role in catalyzing passage through the restriction point remains to be determined. These proteins have been identified in three different ways: by cloning dominant oncogenes, by cloning delayed response genes, and by cloning mammalian proteins that can act as G1 cyclins in budding yeast. These approaches have identified four new classes of putative G1 cyclins: cyclins C, D, E, and F. The best candidates for delayed response genes whose products push mammalian cells through the restriction point are members of the cyclin D and cyclin E families. The cyclin D genes are a small family of closely related genes, and one member has been identified as an oncogene in some human tumors. Although the transcription of cyclin D genes is induced by stimulating quiescent cells with growth factors (Figure 6-22), no marked fluctuation occurs in the abundance of the mRNA or protein as exponentially growing cells traverse the cell cycle.

The strongest suggestion that mammalian G1 cyclins regulate passage through the restriction point comes from overexpressing the cyclin E gene. Normally the abundance of cyclin E rises and falls in the mammalian cell cycle in much the same way as that of the G1 cyclins in yeast. In cell lines in which the cyclin E gene is expressed from a constitutive promoter, the duration of G1 and the minimum serum concentration that cells need to proliferate are reduced. Thus the effects of increasing the level of cyclin E in mammalian cells resemble the way in which elevated levels of Cln3 decrease the duration of G1 in budding yeast and make it more difficult for mating factors to arrest the cell cycle.

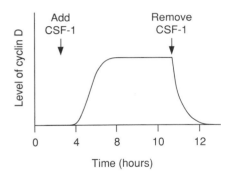

FIGURE 6-22 Cyclin D expression. Synthesis of cyclin D mRNA and protein is induced by the macrophage growth factor, colony-stimulating factor 1 (CSF-1). When the growth factor is removed, the level of cyclin D falls rapidly. Reference: Matsushima, H., Roussel, M. F., Ashmun, R. A., & Sherr, C. J. *Cell* **65**, 701–713 (1991).

Relatives of Cdc2/28 probably drive mammalian cells through the restriction point

In both budding and fission yeast, Cdc2/28 plays a crucial role at Start and at mitosis. The same is not true in mammalian cells, in which the role of Cdc2/28 appears to be exclusively mitotic. Temperature-sensitive mutations in mouse Cdc2/28 prevent cells from entering mitosis but allow them to pass the restriction point. Passage through the restriction point is likely to be stimulated by complexes between G1 cyclins and close relatives of Cdc2/28. These proteins are referred to as **cyclin-dependent kinase (Cdk) proteins,** and one of them, Cdk2, has been identified in a wide variety of vertebrates and can form complexes with cyclin E and cyclin A. Cdk2 is a delayed response protein: It is present at a constant level in exponentially growing cells but disappears rapidly as cells are starved and enter G0. The behavior of G1 cyclins and Cdk2 suggests that starvation not only arrests, but actually dismantles, the part of the cell cycle engine that drives cells through the restric-

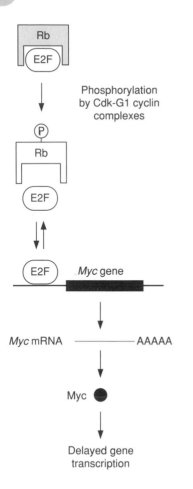

Phosphorylation by Cdk-G1 cyclin complexes

Myc gene

Myc mRNA ——————— AAAAA

Myc ●

Delayed gene transcription

FIGURE 6-23   The retinoblastoma protein, a cell cycle brake. The transcription factor E2F activates the expression of some early response genes such as *c-myc,* which encodes a transcription factor. During G1, E2F is bound to the tumor suppressor Rb, preventing E2F from activating *c-myc* transcription. At the restriction point Rb is phosphorylated by Cdk–cyclin complexes, causing a conformational change in the protein, which releases E2F and allows it to bind to DNA and activate transcription. Reference: Chellappan, S. P., Hiebert, S., Mudryj, M., Horowitz, J. M., & Nevins, J. R. *Cell* **65,** 1053–1061 (1991); Ludlow, J. W., Shon, J., Pipas, J. M., Livingston, D. M., & DeCaprio, J. A. *Cell* **60,** 387–396 (1990).

tion point. Early response proteins presumably induce the transcription of G1 cyclins and Cdk2, but there is not yet direct evidence to support this hypothesis. Eukaryotes probably contain many different protein kinases that are complexes of Cdk and cyclin-like proteins, and some of these may regulate processes other than the cell cycle.

Cdk–G1 cyclin complexes release the brake on progress through the restriction point

To understand how cellular proliferation is controlled, we need to know how transcription of early response genes is regulated. Most of our information on this topic comes from studying a transcription factor called **E2F** that appears

FIGURE 6-24   Positive feedback loop for G1 cyclin expression in mammalian cells. Passage through the restriction point may involve an autocatalytic loop like the one that drives budding yeast cells through Start. Because Cdk–G1 cyclin complexes can phosphorylate and inactivate the Rb protein, weak expression of the G1 cyclins inhibits Rb function, leading to increased G1 cyclin expression and further inactivation of Rb. Eventually the protein kinase activity of the Cdk–cyclin complexes is thought to reach a level that induces the initiation of DNA replication.

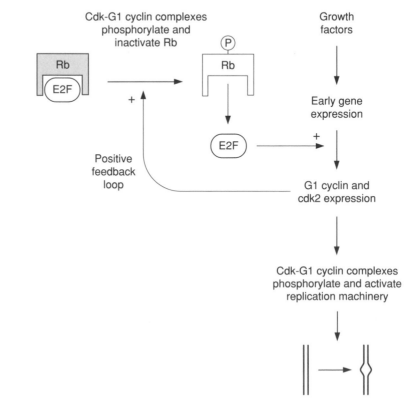

to play a role in inducing the transcription of both early and delayed response genes. During G0 and G1 the products of two tumor suppressor genes, the **retinoblastoma protein (Rb)** and a close relative called p107, act as brakes on cell cycle progress by binding to E2F. The Rb – E2F complex cannot stimulate transcription, providing a partial explanation of how brake proteins prevent passage through the restriction point (Figure 6-23).

How do growth factors and the signals they induce antagonize the ability of tumor supressor gene products like Rb to inhibit the transcription of early response genes? An important clue comes from studying the phosphorylation of the brake proteins during the cell cycle. Rb is dephosphorylated as cells leave mitosis and phosphorylated again late in G1, at a time that corresponds roughly to the restriction point. In vitro these phosphorylations can be induced by Cdc2/28 – cyclin B, Cdk2 – cyclin A, or Cdk2 – cyclin E complexes.

Phosphorylation of Rb blocks its ability to bind to E2F. Thus, like their role in budding yeast at Start, the action of the G1 cyclins at the restriction point in mammalian cells may be autocatalytic (Figure 6-24). The ability of G1 cyclin-containing complexes to inactivate brake proteins like Rb and p107 would increase the transcription of early response genes, which would in turn stimulate the transcription of G1 cyclins and other delayed response genes. Thus, small amounts of G1 cyclins could trigger explosive increases in the rate of G1 cyclin transcription.

CONCLUSION

The basic reactions that control cell proliferation in budding yeast and in mammalian cells are remarkably similar. In budding yeast the association of G1 cyclins with Cdc28 produces an active protein kinase that induces passage through Start, and in mammalian cells the evidence is growing that G1 cyclins associate with close relatives of Cdc2/28 to induce passage through the restriction point. Analyzing how mating factors arrest the cell cycle of budding yeast has provided a detailed description of a signaling pathway that regulates the cell cycle by modulating the levels of G1 cyclins in response to signals generated by other cells. The striking similarities between this protein kinase cascade and the one that induces cell proliferation in mammalian cells make two important points: The proteins that control the cell cycle engine have been conserved during evolution, but the same pathway can stimulate proliferation in one cell type and inhibit it in another. The analysis of Ras function in yeast and mammalian cells emphasizes the danger of assuming that the principal function of an evolutionarily conserved protein is the same in two different cells.

The control of proliferation in budding yeast is fairly well understood, but despite prodigious labors, we have many more questions than answers about control of growth and proliferation in mammalian cells. Do G1 cyclins play the same role in mammalian cells that they do in yeast? What is the relative contribution of cyclin A, whose abundance peaks in S phase, and cyclins D and E in driving cells through the restriction point and inducing DNA replication? Is there really a single restriction point or is there a series of coupled microtransitions in the cell cycle engine? For instance, the early response genes might induce the production of Cdk2 – G1 cyclin complexes, which in turn could induce the expression of cyclin A and the activation of the DNA replication machinery. How many steps in the pathway that leads to the restriction point are inhibited by brake proteins, and how many of these proteins are there? We expect that the next few years will provide a much more concrete picture of how cell proliferation is controlled in multicellular organisms.

SELECTED READINGS

Reviews

Draetta, G., Ed. Cyclin dependent kinases. *Seminars in Cell Biol.* **2**, 193 – 271 (1991). *A series of reviews on the biochemistry and biology of Cdc2 – cyclin complexes.*

Hunt, T., & Kirschner, M. W., Eds. Cell multiplication. *Curr. Opinion in Cell Biol.* **5**, Issue 2 (1993). *A series of reviews on various aspects of the cell cycle. This publication reviews advances in the cell cycle each year.*

Marsh, L., Neiman, A. R., & Herskowitz, I. Signal transduction during pheromone response in yeast. *Ann. Rev. Cell Biol.* **7**, 699 – 728 (1991). *A review of the signaling pathway between mating factors and the cell cycle engine in budding yeast.*

Heldin, C. H. Structural and functional studies on platelet derived growth factor. *EMBO J.* **11**, 4251 – 4259 (1992). *A review of the discovery of PDGF and its actions on cells.*

Bourne, H. R., Sanders, D. A., & McCormick, F. The GTPase superfamily: A conserved switch for diverse cell functions. *Nature* **348**, 125 – 132 (1990).

Bourne, H. R., Sanders, D. A., & McCormick, F. The GTPase superfamily: Conserved structure and molecular mechanism. *Nature* **349**, 117 – 127 (1991). *A pair of reviews that discuss the structure and function of G proteins and the molecules that interact with them.*

Sherr, C. J. Mammalian G-1 cyclins. *Cell* **73**, in press (1993). *A review of the discovery of the G-1 cyclins and their role in the mammalian cell cycle.*

Original Articles

Cross, F. R. *DAF1,* a mutant gene affecting size control, pheromone arrest, and cell cycle kinetics of *Saccharomyces cerevisiae. Mol. Cell. Biol.* **8**, 4675 – 4684 (1988). *Discovery of the first G1 cyclin in budding yeast.*

Richardson, H. E., Wittenberg, C. W., Cross, F., & Reed, S. I. An essential G1 function for cyclin-like proteins in yeast. *Cell* **59**, 1127 – 1133 (1989). *A demonstration that the three G1 cyclins in yeast are redundant and made and destroyed in each cell cycle.*

Broek, D., *et al.* The *S. cerevisiae* CDC25 gene product regulates the RAS/adenylate cyclase pathway. *Cell* **48**, 789 – 799 (1987). *The demonstration that the budding yeast Cdc25 gene acts through Ras to control the level of cyclic AMP.*

Matsushime, H., Roussel, M. F., Ashmun, R. A., & Sherr, C. J. Colony-stimulating factor 1 regulates novel cyclins during the G1 phase of the cell cycle. *Cell* **65**, 701 – 713 (1991). *The discovery that cyclin D is a delayed response gene in mammalian cells.*

Chellappan, S. P., Hiebert, S., Mudryj, M., Horowitz, J. M., & Nevins, J. R. The E2F transcription factor is a cellular target for the RB protein. *Cell* **65**, 1053 – 61 (1991). *The demonstration that the Rb protein binds to and sequesters a factor involved in the transcription of early and delayed response genes.*

THE CONTROL OF DNA
REPLICATION

CELLS MUST replicate their DNA and microtubule organizing centers in each cell cycle. How does passage through Start or the restriction point induce replication of these components? How do cells make sure that they replicate their DNA only once in each cell cycle? This chapter describes experiments that have suggested answers to these questions.

At mitosis, MPF induces downstream events by directly phosphorylating structural and regulatory proteins that control the architecture of the cell (see Chapter 5). At Start (in yeasts), active Cdc2/28 – G1 cyclin complexes induce the replication of DNA and the microtubule organizing center; at the restriction point (in mammalian cells) the kinases that induce these downstream events are probably Cdk – G1 cyclin and Cdk – cyclin A complexes. Although we are beginning to understand how cells regulate passage through Start or the restriction point (see Chapter 6), we know little about how passage through these checkpoints induces downstream events. What proteins are modified to induce DNA replication and duplication of the microtubule organizing center? Are these changes directly or indirectly induced by Cdc2/28 – or Cdk – G1 cyclin complexes? Since complexes of cyclins and Cdk molecules appear to induce both Start and mitosis, how do yeast cells ensure that Cdc2/28 – G1 cyclin complexes induce the downstream events of Start, while Cdc2/28 – cyclin B complexes induce the downstream events of mitosis?

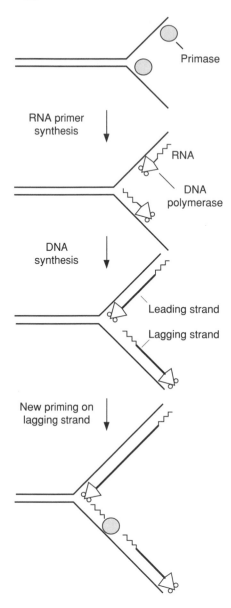

FIGURE 7-1 Basic principles of DNA replication. DNA polymerases have two important properties that affect replication: They can polymerize DNA in only one direction, and they must begin by adding nucleotides to a preexisting primer. Because of these restrictions, small RNA molecules are synthesized to act as primers for DNA synthesis. On the leading strand, after synthesis of the initial primer, the polymerase can continue for thousands of bases. By contrast, on the lagging strand new primers must be made frequently to allow sequences exposed by the progress of the leading strand to be replicated.

ACTIVATION OF THE REPLICATION MACHINERY

To replicate its DNA, a cell must perform two tasks: It must activate the **replication origins,** the specialized regions of the chromosome where DNA replication starts, and it must activate the enzymatic machinery that replicates the DNA. Once replication has been initiated at an origin (called firing the origin), **replication forks** move away from the origin in either direction.

In all organisms DNA is replicated by **DNA polymerases,** which use a single-stranded DNA molecule as a template to direct the synthesis of the complementary strand. The polymerases have two important properties that affect the mechanism of DNA synthesis: They cannot start a new nucleotide chain, and they can synthesize DNA chains running in only one direction. The inability of DNA polymerases to start new chains means that they require a nucleotide chain that they can extend, called a **primer,** as well as a template. During chromosomal DNA replication, the primers are very short RNA chains that are synthesized by specialized RNA polymerases called **primases.** In the final stages of DNA replication, the primers are removed and replaced with DNA with the result that at the end of replication, each chromosome is composed of two continuous strands of DNA.

Since the two nucleotide chains in a DNA molecule run in opposite directions, the two newly synthesized strands at the replication fork also run in opposite directions. One new strand, the **leading strand,** runs in the same direction as the DNA polymerase and is synthesized continuously. The other new strand, the **lagging strand,** runs in the opposite direction to the DNA polymerase and must be synthesized in short segments that are later joined together (Figure 7-1).

DNA viruses are model systems for studying DNA replication

In the 1960s experiments on bacterial cells showed that DNA replication begins at a specific point on the chromosome called the replication origin (see pages 190–193). Unlike bacteria, eukaryotes have multiple replication origins on each chromosome, complicating studies on the initiation of DNA synthesis. Many of these difficulties can be avoided by studying the replication of certain animal viruses, whose genome is composed of double-stranded DNA that has a single replication origin. These viruses use the host cell's enzymatic machinery to replicate their DNA, with the single exception that the virus encodes a DNA-binding protein that recognizes its own replication origin, rather than those of the cell it is infecting.

The best-studied replication origin belongs to a small DNA virus called **SV40** (simian virus 40), which was originally discovered as a contaminant of the earliest preparations of polio vaccine. Replication from this origin requires a virally encoded protein, called **large T,** and several proteins provided by the cell in which the virus is replicating. All of these proteins have been purified and can be mixed together in a test tube to replicate SV40 DNA. The binding of large T to the three tandemly repeated large T binding sites that form the SV40 replication origin is the first step in initiating replication (Figure 7-2). Large T is a **DNA helicase:** it hydrolyzes ATP and uses the energy produced to separate the two DNA strands and move itself along the DNA, progressively unwinding it. An additional protein, called **RF-A** (replication factor), binds to the unwound single strands to keep them from reannealing. The combined action of large T and RF-A converts the replication origin into single stranded DNA, allowing a complex of DNA polymerase and primase to bind to the opened origin. The primase synthesizes an RNA primer that the

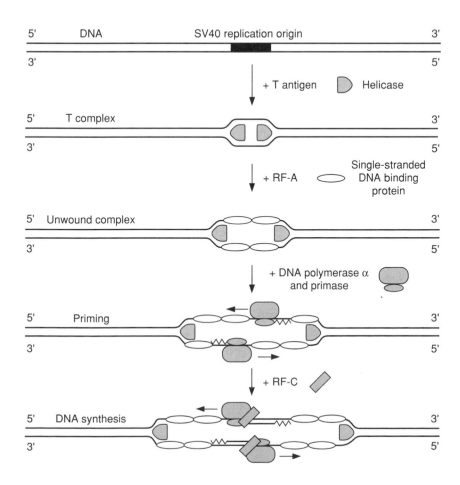

FIGURE 7-2 SV40 DNA replication. Small DNA viruses are excellent systems in which to study the biochemistry of eukaryotic DNA replication. SV40 (simian virus 40) encodes large T antigen, a protein that binds to its replication origin, but otherwise relies on DNA replication proteins that are already present in the host cell. The binding of T antigen to the origin and its DNA helicase activity lead to the unwinding of a region around the origin. Subsequent binding of a primase allows the RNA primers that the DNA polymerases will extend to be produced. After the initiation of DNA replication, complex changes occur to assemble specific complexes of polymerases and accessory factors on the leading and lagging strands. References: Dutta, A., & Stillman, B. *EMBO J.* 11, 2189–2199 (1992); Erdile, L. F., et al. *Cold Spring Harbor Symp. Quant. Biol.* 56, 303–313 (1991).

polymerase extends to form the leading strand. The details of DNA elongation are complex. There appear to be separate DNA polymerases for the leading and lagging strands, and the leading strand polymerase has **accessory proteins** that allow it to polymerize many thousands of nucleotides without ever falling off the template.

DNA synthesis in budding yeast begins at specific replication origins

How is chromosomal DNA replicated? Because all the proteins involved in SV40 replication, with the exception of large T, are cellular proteins, chromosomal DNA replication is likely to be very similar to that of SV40. Instead of large T binding to the SV40 replication origin, we expect that the binding of a cellular origin-binding protein to chromosomal replication origins will initiate cellular DNA replication. The first step in understanding chromosomal replication was the identification of replication origins in budding yeast, which was accomplished by finding pieces of yeast DNA that would allow a small circular DNA molecule to replicate in the nucleus of a yeast cell. The *ARS* (autonomously replicating sequences) elements are located about 50,000 base pairs apart on budding yeast chromosomes and have been demonstrated to be the points at which replication initiates (Figure 7-3). Each *ARS* consists of a strongly conserved adenine/thymine-rich DNA sequence that is flanked by other, less well conserved elements.

Recently, considerable progress has been made toward identifying origin-specific binding proteins that could regulate DNA replication. Purification of a factor that binds to *ARS* sequences has revealed a large complex of six proteins. The binding of the complex to DNA requires ATP, but the complex does not have the type of helicase activity that large T uses to unwind the SV40 DNA replication origin. We do not yet know if the binding of this complex to DNA triggers replication, or whether the complex remains bound throughout the cell cycle and is activated in S phase by post-translational modification or association with other proteins.

How do embryonic cells replicate their DNA so much more rapidly than their somatic counterparts? One reason is that the replication machinery in embryonic cells is active throughout the cell cycle (page 126). Another is that although replication forks in early embryonic and somatic cells move at the same speed, early embryonic cells have many more active replication origins

FIGURE 7-3 Budding yeast replication origins. Budding yeast replication origins were discovered as sequences that allow the plasmids that carry them to replicate autonomously inside the nucleus of a yeast cell. These *ARS* elements share a common structure, (bottom) consisting of a highly conserved A/T-rich sequence (A) flanked by more weakly conserved, but nevertheless important, sequences (B1, B2, B3).

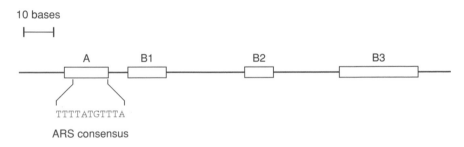

than somatic cells. As a result the forks do not have to travel as far in each cell cycle. Embryos apparently achieve this numerical advantage by allowing any sequence in their DNA to act as a replication origin, instead of limiting initiation to specific DNA sequences.

Cells activate their replication machinery as they enter S phase

Cell fusion experiments first showed that the ability of cells to replicate DNA varies during the cell cycle. The ability of mammalian cell extracts to replicate DNA also varies, depending on the stage of the cell cycle from which the extract was prepared. The activity of extracts was assessed by mixing them with SV40 DNA and large T. Extracts made from G1 cells were much less competent to replicate SV40 DNA than extracts made from S phase cells. Mixing S phase and G1 extracts produced levels of SV40 replication similar to those in S phase extracts, suggesting that the S phase extracts contained a factor required for DNA replication that was absent from, or inactive in, the G1 extract. Partial purification revealed that this factor was RF-A, the protein that helps T antigen unwind the viral replication origin. The S phase extracts also contained an activity that could turn on the inactive RF-A in G1 extracts. Partial purification suggested that this factor was a complex of a Cdk protein and a cyclin, and in vitro both Cdc2/28 – cyclin B and Cdk2 – cyclin A complexes can phosphorylate RF-A.

There are many unanswered questions about the control of DNA replication. Which Cdk – cyclin complexes are responsible for phosphorylating and activating RF-A in vivo? Is this the only component of the replication machinery whose activity is regulated during the cell cycle? Since large T is phosphorylated and activated by Cdk – cyclin complexes, it seems likely that the protein complexes that bind cellular replication origins are similarly regulated. In the control of DNA replication, what are the relative roles of changing the amount of replication proteins and post-translationally altering the activity of preexisting proteins? This question has no clear answer. In growing fission yeast cells, most of the proteins involved in DNA replication are made throughout the cell cycle, but in both budding yeast and mammalian cells many of the functions involved in DNA replication are expressed periodically during the cell cycle. In budding yeast, engineering strains to express these proteins throughout the cell cycle does not affect the timing of DNA replication, suggesting that periodic gene expression alone is not responsible for the existence of a discrete S phase.

MICROTUBULE ORGANIZING CENTER DUPLICATION

Duplication of the microtubule organizing center is an important downstream event of Start or the restriction point. During mitosis the microtubule organizing center forms the poles of the spindle, so that each newborn cell inherits one microtubule organizing center, which must duplicate before the next mitosis. In budding yeast the duplication of the microtubule organizing center is clearly induced at Start. In other organisms the situation is less clear cut. In vertebrate cells, duplication of the centrioles, which lie at the core of the microtubule organizing center, is induced when cells pass through the restriction point, but two functionally distinct microtubule organizing centers do not appear until late in G2. In fission yeast, duplication of the microtubule organizing center seems to occur in G2, but this might reflect an earlier Start-induced event analogous to centriole duplication in vertebrate cells.

Spindle pole body duplication occurs at Start in budding yeast

In budding yeast the spindle pole body is the microtubule organizing center. The spindle pole body is intimately associated with the nuclear membrane and nucleates cytoplasmic and nuclear microtubules from its opposite faces. Cells are born with a single spindle pole body, which enlarges at Start and then splits into two daughter spindle pole bodies, which are briefly connected by a bridge before separating to opposite sides of the nucleus (Figure 7-4). At the restrictive temperature $cdc28^{ts}$ mutants fail to enlarge or duplicate their spindle pole body, suggesting that Cdc28 activity induces spindle pole body duplication. The earliest steps in spindle pole body duplication actually occur in the previous cell cycle, since the functions of some genes are required during the previous cell cycle to make the spindle pole body competent to duplicate. Other genes act at or after Start on the pathway of spindle pole body duplication.

Centrosomes have an autonomous replication cycle

In animal cells, the core of the centrosome is a pair of **centrioles.** The two centrioles, each a hollow cylinder composed of 27 specialized microtubules, lie at right angles to each other. During the cell cycle each centriole gives rise to a new centriole. At prophase the two pairs of centrioles each organize a functional centrosome, and the two centrosomes migrate apart from each

FIGURE 7-4 Spindle pole body replication in budding yeast. The spindle pole body appears to replicate in discrete stages, which have been visualized by electron microscopy. Only those microtubles that lie on the nuclear side of the spindle pole body are shown. The genes required for different stages of duplication are indicated. Reference: Winey, M., Goetsch, L., Baum, P., & Byers, B. *J. Cell Biol.* **114**, 745– 754 (1991).

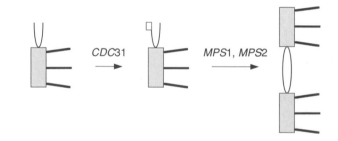

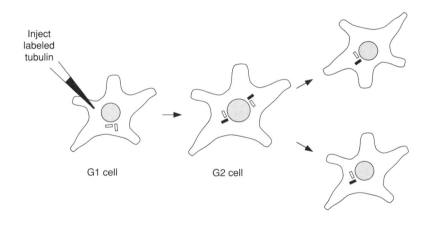

Inject
labeled
tubulin

G1 cell   G2 cell

FIGURE 7-5 Centriole replication. The mode of centriole replication in mammalian cells was investigated by injecting G1 cells with chemically labeled tubulin molecules. After the centrioles had replicated and the cells had divided, the cells were fixed, and the chemical tag was visualized in the daughter cells by using antibodies directed against the tag. Each daughter cell contained one labeled and one unlabeled centriole, showing that material is not mixed between the old and new centriole during replication. Reference: Kochanski, R. S., & Borisy, G. G. *J. Cell Biol.* **110,** 1599–1605 (1990).

other to form the two spindle poles. Although centrioles are associated with most centrosomes, several different cell types, such as early mammalian embryos, lack centrioles.

The centriole is poorly understood. New centrioles normally arise only by the duplication of existing centrioles, but in some meiotic cell cycles centrioles form in the absence of any other centrioles. Unlike DNA, which replicates semiconservatively, with each daughter double helix receiving one strand from the parental double helix, the centriole replicates conservatively, with an entirely new centriole forming alongside an old one (Figure 7-5). Since the spindle pole body and centrioles, like chromosomes, replicate only once per cell cycle, speculation that they contain a nucleic acid component has persisted. The most careful experiments, however, have failed to demonstrate any nucleic acid component required for microtubule organizing center function or duplication.

In somatic cells centrosome and centriole replication is tightly coupled with the cell cycle: The centrosomes and centrioles each replicate only once per cycle. In embryonic cell cycles, however, the coupling between the centrosome replication cycle and the cell cycle is not so rigid. Treating early embryos with protein synthesis inhibitors arrests the cell cycle in interphase, but centriole and centrosome duplication continues. This experiment suggests that there is a separate centrosome cycle that normally is entrained by the cell cycle engine but continues to run when the engine is halted.

Perhaps, like the DNA replication machinery, the centrosome duplication machinery is active throughout early embryonic cell cycles. In a sense we could regard the embryonic cells as being at Start throughout the cycle. If this view is correct, how do embryonic cells ensure that they replicate their centrosomes only once per cell cycle? One hypothesis is that the act of replication makes a centrosome incapable of duplicating again for a period of time. Passage through mitosis would lift the block and allow centrosome duplication to occur in the next cell cycle. As long as the block to reduplication were longer than the length of the cell cycle, the centrosome would duplicate only once per cycle. If the block were not absolute, however, it would gradually decay in interphase-arrested cells, allowing another round of centrosome duplication to occur.

Analyzing the budding of *Saccharomyces cerevisiae* also suggests the idea of blocks that prevent a given event from occurring more than once per cycle. Normal cells bud only once per cell cycle, but *cdc4^ts,* and *cdc34^ts* mutants have the unusual property that they bud repeatedly at the restrictive temperature, without replicating their DNA or performing mitosis. Both of these mutants arrest with high levels of kinase activity associated with Cdc28–G1 cyclin, suggesting they arrest in Start. One explanation of the multiple budding phenotype is that cells can bud only at Start and that the act of budding produces an inhibitory signal that keeps them from budding again until they have gone through mitosis. If the inhibitory signal decayed with time, cells arrested in Start would eventually bud again.

Cdc34 is an enzyme that conjugates ubiquitin to proteins, thus targeting them for proteolysis. Since *cdc34^ts* arrests in Start, it seems likely that some proteins must be degraded so that cells can pass through Start and begin DNA replication. What are these proteins? G1 cyclins are stable in *cdc34^ts* strains at the restrictive temperature, suggesting that they are substrates for Cdc34. But since mutations in the G1 cyclins that render them stable do not arrest cells in Start, we suspect that there are other proteins whose degradation is required for progression into S phase.

THE BLOCK TO REREPLICATION

Mammalian cell fusion experiments first suggested that cells have mechanisms that keep them from replicating their DNA twice in a single cell cycle (see pages 12–14). How does this block to rereplication work, and how is it removed by passage through mitosis so that rounds of chromosome replication and chromosome segregation alternate with each other? As with many other "rules" of the cell cycle, the rule that cells replicate and segregate their chromosomes alternately has exceptions. Meiotic cell cycles (see Chapter 10) consist of a single round of replication followed by two successive rounds of chromosome segregation; **endoreduplication** cycles produce highly polyploid cells, such as those found in *Drosophila* embryos, by having repeated rounds of DNA replication without any intervening chromosome segregation (Figure 7-6). The existence of these specialized cell cycles demonstrates that the principle of alternating replication and segregation is the result of checks and balances built into the cell cycle, rather than an inevitable consequence of the fundamental organization of the cycle.

DNA replicates only once per cell cycle

Cells cannot segregate their genetic information accurately if regions of the chromosome fail to replicate, or replicate more than once per cell cycle. Failure to replicate part of a chromosome produces two sister chromatids

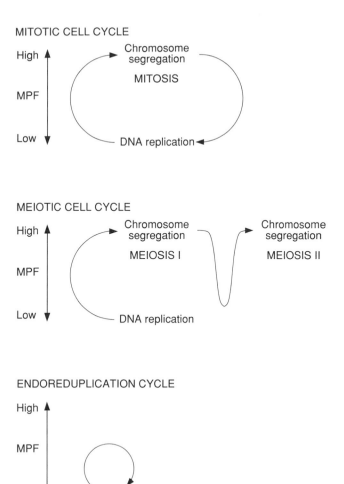

MITOTIC CELL CYCLE

MEIOTIC CELL CYCLE

ENDOREDUPLICATION CYCLE

FIGURE 7-6 FIGURE 7-6 *Different cell cycles. In normal cell cycles rounds of chromosome segregation and DNA replication alternate. In the meiotic cell cycle, a single round of replication is followed by two rounds of chromosome segregation. In endoreduplication cycles, successive rounds of DNA replication occur without intervening rounds of chromosome segregation.*

attached to each other by a region of unreplicated DNA. Trying to segregate these chromosomes either breaks one of the sisters or causes them both to segregate to one pole of the spindle **(nondisjunction)** (Figure 7-7). If part of a chromosome replicates twice, this region forms a replication bubble on the chromosome. If the bubble encompasses the centromere, the chromosome will have two centromeres and will be broken apart or mis-segregated at mitosis (Figure 7-8).

Because each of their chromosomes contains many replication origins that fire asynchronously, eukaryotic cells need special mechanisms to ensure that all their DNA is replicated just once in each cycle. If a replication fork passes over an unreplicated origin, the origin must be inactivated immediately to keep that part of the chromosome from replicating twice (Figure 7-9). Therefore, the block to rereplication must ensure not only that no origin fires more than once in any cell cycle, but also that any origin that has not fired is rapidly inactivated when traversed by a replication fork. This consideration suggests that controlling the activity of individual replication origins, rather than that of the replication machinery as a whole, is the key to preventing rereplication during a single cell cycle.

FIGURE 7-7 Incomplete DNA replication damages chromosomes . Failure to replicate part of a chromosome produces two sister chromatids that are irreversibly linked. At anaphase either the chromosomes break at this linkage point, or one of the two kinetochores becomes detached so that both sisters arrive at one pole (nondisjunction).

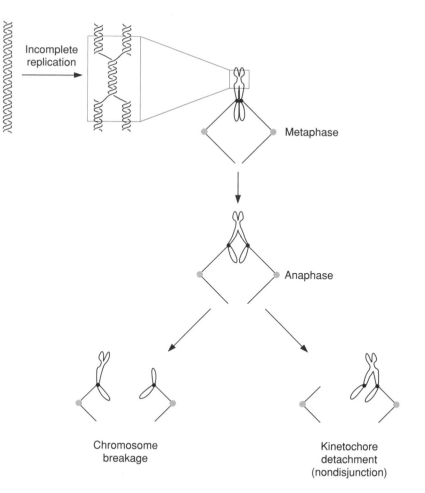

Metaphase

Anaphase

Chromosome breakage

Kinetochore detachment (nondisjunction)

Embryonic cells remove the block to rereplication at the end of mitosis

Like somatic cells, embryonic cells replicate their DNA only once per cell cycle, even though the replication machinery is continually active. In early embryos there is no pause between the exit from mitosis and the beginning of DNA replication: As soon as chromosomes decondense and acquire nuclear envelopes, they begin to replicate their DNA. Is the replication machinery inactivated when chromosomal DNA synthesis finishes? This question was addressed by injecting foreign DNA into fertilized frog eggs which had been arrested in interphase by inhibiting protein synthesis. Even if it was injected long after the chromosomal DNA had completed replication, the foreign DNA formed nuclei and replicated (Figure 7-10) demonstrating that the replication machinery remains active throughout interphase.

The temporal correlation between the exit from mitosis and the onset of DNA replication suggests that the block to rereplication is removed as a result of passage through mitosis. This idea was confirmed by inducing interphase-arrested eggs to pass through mitosis by injecting them with MPF. The injection of MPF drove the eggs into mitosis, but as MPF levels declined, the eggs entered the next interphase and rereplicated their DNA (see Figure 7-10).

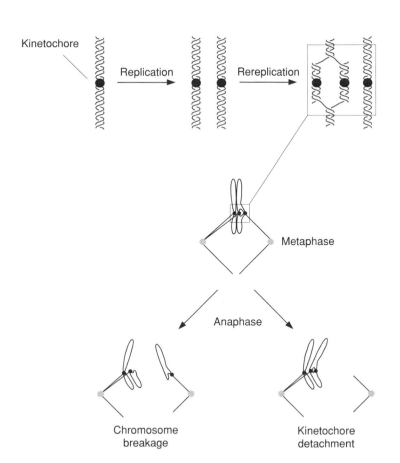

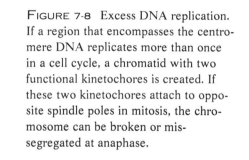

FIGURE 7-8 Excess DNA replication. If a region that encompasses the centromere DNA replicates more than once in a cell cycle, a chromatid with two functional kinetochores is created. If these two kinetochores attach to opposite spindle poles in mitosis, the chromosome can be broken or missegregated at anaphase.

DNA replication is controlled by a licensing factor that enters the nucleus at mitosis

How does passage through mitosis release the block to rereplication? Recent experiments on frog eggs suggest that the ability to replicate DNA is restored by the physical act of breaking down the nuclear envelope (Figure 7-11). Adding sperm nuclei to interphase arrested frog cell cycle extracts induces a single round of DNA replication. If the replicated nuclei are removed from the extract and added to a fresh interphase extract, they do not rereplicate their DNA. But if the isolated nuclei are treated with agents that make the nuclear envelope permeable before adding them to the second extract, they do replicate their DNA again. Resealing the nuclei before adding them to the second extract blocks their ability to rereplicate, demonstrating that rereplication requires an interaction between components in the cytoplasm and the contents of the nucleus. In addition, rereplication does not occur unless the second extract contains vesicles that can fuse with the added nuclei to reseal the holes.

These experiments led to the **licensing factor model** for control of DNA replication (Figure 7-12). The model proposes that DNA replication is con-

FIGURE 7-9 Replication forks inactivate replication origins. In chromosomes with multiple origins that fire asynchronously, replication forks inactivate unfired origins as they pass over them. Without this mechanism, the firing of an origin that has already been traversed by a replication fork causes rereplication of part of the chromosome.

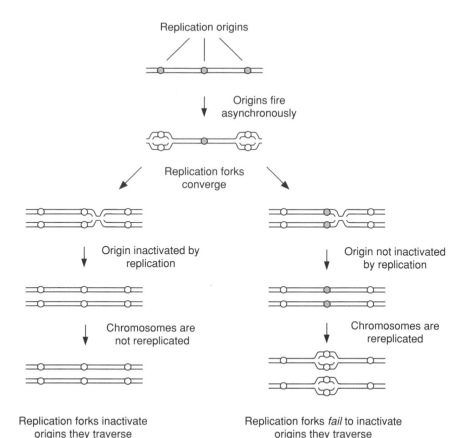

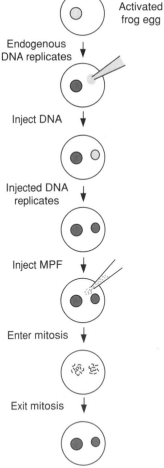

trolled by a **licensing factor** that binds to the chromosomes during mitosis. Even when licensing factor has bound to the chromosomes, replication cannot begin until they are enclosed in a complete nuclear envelope, thus ensuring that DNA replication does not begin during mitosis. To explain the block to rereplication, the model proposes that the licensing factor that is inside the nucleus is consumed or inactivated as the DNA replicates. When the nuclei break down at the onset of mitosis, more licensing factor can bind to the chromosomes, enabling them to replicate in the next interphase.

FIGURE 7-10 DNA replication in early embryos. A fertilized frog egg arrested in interphase by treatment with a protein synthesis inhibitor will replicate its nuclear DNA only once. DNA injected into the egg forms an artificial nucleus and also replicates only once, even if injection occurs long after the endogenous DNA has been replicated. Injection of MPF induces nuclear envelope breakdown in such an egg, and when the nuclei re-form, they replicate their DNA again (heavy stippling). Reference: Newport, J. W., & Kirschner, M. W. *Cell* **37**, 731–742 (1984).

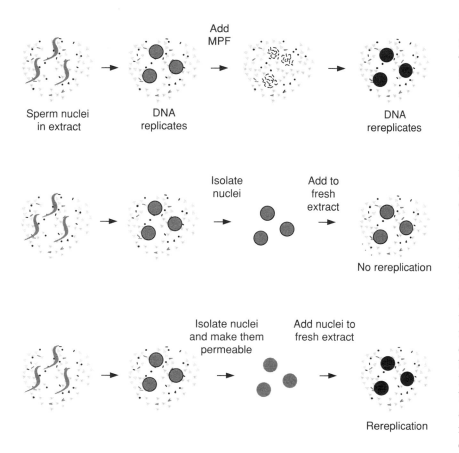

Sperm nuclei
in extract

DNA
replicates

Add
MPF

DNA
rereplicates

Isolate
nuclei

Add to
fresh
extract

No rereplication

Isolate nuclei
and make them
permeable

Add nuclei to
fresh extract

Rereplication

FIGURE 7-11 Nuclear envelope breakdown controls DNA replication. Three different experiments performed in extracts from frog eggs are shown. In one (top) sperm were added to an extract arrested in interphase and replicated their DNA once (light stippling). Adding purified MPF to the extract induced nuclear envelope breakdown, followed by re-formation of the nuclei and a second round of replication (heavy stippling). In the second experiment (middle) sperm nuclei were allowed to replicate and then were isolated from the extract and added to a fresh extract. This manipulation did not induce rereplication. In the third experiment (bottom) the nuclei were allowed to replicate and then were isolated and made permeable with a membrane-disrupting agent. Adding these nuclei back to a fresh interphase extract did lead to rereplication, showing that breaking the barrier between nucleus and cytoplasm is sufficient to induce a new round of DNA replication. Reference: Blow, J. J., & Laskey, R. A. *Nature* **332,** 546–548 (1988).

The licensing factor model explains how chromosome replication and segregation alternate with each other in early embryonic cell cycles. Since licensing factor can gain access to the DNA only during mitosis, two rounds of DNA replication cannot occur without an intervening round of mitosis and chromosome segregation. Two rounds of mitosis cannot occur without an intervening round of replication because the inactivation of MPF leads rapidly to the onset of replication, whereas the induction of the next round of mitosis requires the slow steps of accumulating cyclin and converting preMPF into active MPF.

Mating-type switching in budding yeast provides an analogy for the control of DNA replication in somatic cells

How is the ability to replicate DNA regulated in somatic cell cycles? We believe that licensing factors bind to DNA during mitosis but cannot stimulate DNA replication until cells have passed through Start. The strongest evidence for this idea comes from studies not of DNA replication but of the phenomenon in budding yeast called mating-type switching (see page 36). In the wild, haploid budding yeast can switch from one mating type to the other as often as once per generation (Figure 7-13). Switching is initiated by an

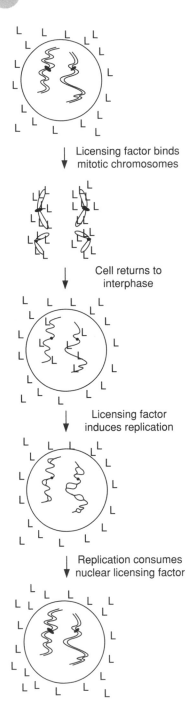

Licensing factor binds
mitotic chromosomes

Cell returns to
interphase

Licensing factor
induces replication

Replication consumes
nuclear licensing factor

FIGURE 7-12   Licensing factor model of DNA replication. When cells enter mitosis, nuclear envelope breakdown allows licensing factor (L) to bind to the chromosomes. Because replication requires an intact nuclear envelope, no immediate replication occurs. When mitosis ends, the re-formation of the nuclear envelope allows the bound licensing factor to induce a round of DNA replication. During replication the licensing factor inside the nucleus is destroyed or inactivated to prevent a second round of DNA replication. The DNA cannot replicate again until a subsequent mitosis allows fresh licensing factor to bind to the chromosomes. Reference: Blow, J. J., & Laskey, R. A. *Nature* **332**, 546–548 (1988).

enzyme that specifically cuts the gene that determines the mating type of the cell. The HO^+ gene (homothallism, the technical term for mating-type switching), which encodes this enzyme, is expressed once per cell cycle, just after passage through Start.

HO^+ expression depends on a transcriptional activator called Swi5 (switch), which acts just like a licensing factor. The cellular location of Swi5 changes during the cell cycle in response to the activity of MPF (Figure 7-14). MPF acts by controlling the activity of the **nuclear localization signal** in the Swi5 protein that allows it to be imported into the nucleus. Swi5 is synthesized during G2, when MPF in budding yeast has already been activated (see page 30). Swi5 remains in the cytoplasm because MPF phosphorylates serines and threonines that flank the nuclear localization signal and thereby blocks the ability of the signal to direct the import of Swi5 into the nucleus. At anaphase the inactivation of MPF leads to dephosphorylation of Swi5, allowing it to enter the nucleus.

How does the entry of Swi5 into the nucleus control HO^+ transcription? Importing Swi5 at the end of closed mitosis parallels the binding of the licensing factors for DNA replication to chromosomes during open mitosis. But unlike DNA replication in early embryonic cell cycles, which begins as soon as the nucleus re-forms, HO^+ transcription does not begin when Swi5 enters the nucleus. Instead, budding yeast cells must pass Start before they can transcribe HO^+ and switch mating types. Dissection of the HO^+ promoter suggests that passage through Start removes or inactivates inhibitory factors that prevent Swi5 from inducing HO^+ transcription. The promoter of the HO^+ gene contains two elements, *URS1* and *URS2*. *URS1* is the site of Swi5 binding; *URS2* appears to inhibit transcription until cells have passed Start (Figure 7-15). In cells in which *URS2* has been deleted, HO^+ expression begins as soon as Swi5 enters the nucleus. The inhibitory effect of *URS2* at Start is overcome by the same Swi4/6 complex that induces the transcription of G1 cyclins (see Chapter 6), but the mechanism involved is unknown.

Start is the point in the cell cycle when the cell allows licensing factors that were imported into the nucleus during mitosis to induce HO^+ transcription. Perhaps DNA replication in the somatic cell cycle and HO^+ transcription are regulated in similar ways. In such a scheme, licensing factors for DNA replication would enter the nucleus at mitosis but would be kept in check by inhibitory activities that would be overcome at Start. In budding yeast the Cdc46 protein, which is a member of a family of related proteins required for DNA replication, has exactly the properties predicted for a licensing factor for DNA replication. The protein is made in G2 but, like Swi5, remains in the cytoplasm until the end of mitosis, when it enters the nucleus.

The idea that Start is a point at which brakes on the cell cycle are released helps unify our views of the embryonic and somatic cell cycles. In both cycles licensing factors for DNA replication can be admitted into the nucleus only at mitosis. In embryonic cells DNA replication begins as soon as the nuclear envelope re-forms because there are no antagonists of the licensing factors. In somatic cells such antagonists exist and prevent replication from beginning as soon as cells enter interphase. The antagonists are inactivated at Start (or the restriction point), producing a temporal gap between the end of mitosis and the beginning of DNA replication. This difference between the embryonic and somatic cell cycles may reflect the different resources that they inherit for DNA replication. Embryos receive large maternal stores of nutrients and replication machinery, but somatic cells must import their nutrients from their environment and produce their own replication machinery. Delaying DNA replication until Start allows somatic cells to take stock of their environment before embarking on the risky task of duplicating their genetic information.

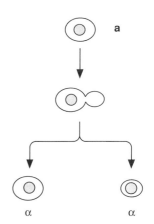

FIGURE 7-13 Mating-type switching in budding yeast. An **a** cell can divide to produce two α cells. Similarly, an α cell can divide to produce two **a** cells (not shown). Only cells that have been a mother at least once can switch mating type. Reference: Herskowitz, I. *Nature* **342,** 749–757 (1989).

Cdc2/28 plays an important role in ensuring the alternation of Start and mitosis

How does the organization of somatic cell cycles ensure that DNA replication alternates with chromosome segregation? Since Start induces DNA replication, and mitosis induces chromosome segregation, Start and mitosis must

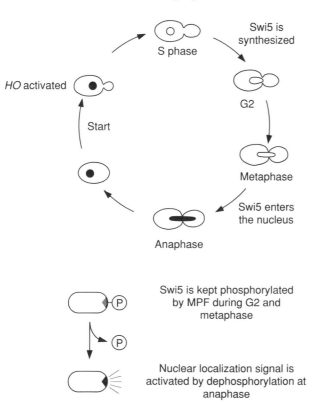

FIGURE 7-14 Swi5, a licensing factor for mating-type switching. Swi5 is a transcription factor that can activate the transcription of the HO^+ gene, whose product is an endonuclease that catalyzes mating type switching. Swi5 is synthesized in G2 but remains in the cytoplasm because phosphorylation by MPF inactivates its nuclear localization signal. At the end of mitosis the inactivation of MPF allows Swi5 to enter the nucleus, but Swi5 cannot stimulate HO^+ transcription until the cell has passed Start. Reference: Moll, T., Tebb, G., Surana, U., Robitsch, H., & Nasmyth, K. *Cell* **66,** 743–758 (1991).

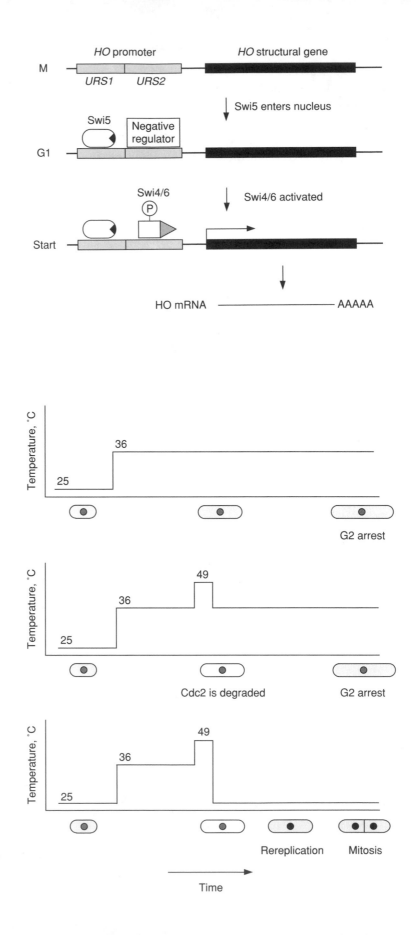

FIGURE 7-15 Activating transcription of HO^+ at Start. The HO promoter has two elements, $URS1$ and $URS2$. After mitosis (M) Swi5 binds to $URS1$ but cannot stimulate transcription because $URS2$ blocks the initiation of transcription. Passage through Start activates the Swi4/6 complex, allowing it to bind to $URS2$ and remove the block to transcription. Reference: Breeden, L., & Nasmyth, K. *Cell* **48**, 389–397 (1987).

FIGURE 7-16 Inducing rereplication in fission yeast. Three manipulations of a $cdc2^{ts}$ fission yeast strain are shown. Shifting cells to 36°C (top) arrests them in G2 with replicated DNA (stippling in nucleus), and Cdc2 protein is still present (stippling in cytoplasm). Shifting the arrested cells briefly to 49°C (middle) induces the degradation of Cdc2, but when the cells are returned to 36°C they do not rereplicate their DNA. If the cells that have been shifted to 49°C are returned to 25°C (bottom) though, newly synthesized Cdc2 is active, and the cells pass Start and rereplicate their DNA (black nucleus). Reference: Broek, D., Bartlett, R., Crawford, K., & Nurse, P. *Nature* **349**, 388–393 (1991).

alternate with each other. An important clue to the mechanism that ensures this alternation came from experiments on special fission yeast *cdc2ᵗˢ* mutants that replicate their DNA twice during the cell cycle. Like most *cdc2ᵗˢ* mutants, these mutants arrested in G2 when an asynchronous population of cells was shifted to 36°C. But if the cells were kept at 36°C for 4 hours, treated at 49°C for 30 minutes, and finally returned to 25°C, they replicated their DNA a second time before entering mitosis (Figure 7-16).

The second round of DNA replication required Cdc2 activity: If the cells were returned to 36°C after the 49°C pulse, they did not rereplicate their DNA, suggesting that they must pass Start before the second round of DNA replication. Only those *cdc2ᵗˢ* alleles that led to the complete destruction of Cdc2 protein during the high-temperature treatment allowed cells to pass Start twice in a single cell cycle and rereplicate their DNA. Alleles of *cdc2ᵗˢ* that inactivated the protein kinase activity of Cdc2 but did not cause its physical destruction did not induce rereplication. This correlation suggests that the Cdc2 protein plays a role in "remembering" whether or not cells have passed Start, but we do not understand the molecular basis of this memory.

The existence of distinct G1 and mitotic cyclins suggests that cells distinguish between Start and mitosis by using different cyclins in these two parts of the cell cycle. Perhaps the cyclins bind to other proteins to localize Cdc2/28-associated kinase activity in the cell, in much the same way that SH2 domains bind proteins that generate second messengers to phosphotyrosine residues on growth factor receptors (see page 107). In keeping with this idea, some proteins are better substrates for Cdk – G1 cyclin complexes than for Cdc2/28 – cyclin B complexes. If this were the only distinction between mitosis and Start, expressing mitotic cyclins in G1 or G1 cyclins in mitosis would be disastrous. Yet yeast strains that express either mitotic or G1 cyclins at high levels throughout the cell cycle grow and divide normally. Therefore, other controls must exist to ensure that Start and mitosis alternate. Possible mechanisms include regulating the ability of different cyclins to associate with Cdc2/28 and Cdk proteins during the cell cycle or cell cycle-specific modifications that regulate the ability of particular proteins to act as substrates for Cdk – cyclin complexes.

CONCLUSION

We believe that we understand the principles that regulate DNA replication and that we are beginning to unearth the molecules that put these principles into practice. Viral DNA replication is understood in considerable detail, and the next few years should see a dramatic increase in our knowledge of the mechanisms that control cellular DNA replication. Compared to this rosy picture, our ignorance about microtubule organizing centers is profound. We do not know their basic composition, how they duplicate, or how they nucleate microtubules. What little we do know about their duplication has revealed a centrosome duplication cycle that is experimentally separable from the cell cycle engine driven by Cdc2/28 – cyclin. Earlier experiments in fission yeast also showed that there are cyclical changes that can occur independently of the cell cycle engine. These include periodic increases in the activity of an enzyme involved in nucleotide metabolism (nucleoside diphosphokinase) and the overall rate of respiration. These observations suggest that there are several different oscillators inside cells, but as far as we know, they are all normally entrained by the cell cycle engine. How many independent oscillators are there, how do they work, and how are they entrained by the cell cycle engine? We have little idea how to approach these important questions, much less answer them.

SELECTED READINGS

General

Kornberg, A., & Baker, T. A. *DNA Replication* (W. H. Freeman, New York, 1992). *A detailed, up-to-date, and comprehensive review of DNA replication in eukaryotes and prokaryotes*

Reviews

Stillman, B. Initiation of eucaryotic DNA replication in vitro. *Ann. Rev. Cell Biol.* **5**, 197–245 (1989).

Diffley, J. F., & Stillman, B. The initiation of chromosomal DNA replication in eucaryotes. *Trends in Genet.* **6**, 427–432 (1990). *Two reviews of the steps that initiate DNA replication. The first is more detailed.*

Karsenti, E., & Maro, B. Centrosomes and spatial organization of microtubules in animal cells. *Trends in Biochem Sci.* **11**, 460–463 (1986). *A concise review of the properties of centrosomes*

Original Articles

D'Urso, G., Marraccino, R. L., Marshak, D. R., & Roberts, J. M. Cell cycle control of DNA replication by a homologue from human cells of the p34cdc2 protein kinase. *Science* **250**, 786–791 (1990). *The first demonstration that Cdc2–cyclin complexes can activate components of the replication machinery.*

Gard, D. L., Hafezi, S., Zhang, T., & Doxsey, S. J. Centrosome duplication continues in cycloheximide-treated Xenopus blastulae in the absence of a detectable cell cycle. *J. Cell Biol.* **110**, 2033–2042 (1990). *A demonstration that centrosomes continue to duplicate when the cell cycle engine is arrested.*

Blow, J. J., & Laskey, R. A. A role for the nuclear envelope in controlling DNA replication within the cell cycle. *Nature* **332**, 546–548 (1988). *The experiments that led to the licensing model factor model for the control of DNA replication.*

Moll, T., Tebb, G., Surana, U., Robitsch, H., & Nasmyth, K. The role of phosphorylation and the CDC28 protein kinase in cell cycle-regulated nuclear import of the *S. cerevisiae* transcription factor SWI5. *Cell* **66**, 743–758 (1991). *Mutations that abolish the ability of Cdc2/28 to phosphorylate Swi5 allow this licensing factor to enter the nucleus constitutively.*

Broek, D., Bartlett, R., Crawford, K., & Nurse, P. p34cdc2 is involved in establishing the dependency of S phase upon completion of the previous mitosis. *Nature* **349**, 388–393 (1991). *The discovery that the physical destruction of Cdc2 allows fission yeast cells to replicate their DNA twice in a cell cycle.*

CHECKPOINTS AND FEEDBACK CONTROLS

THE SURVIVAL of all organisms depends on the accurate transmission of genetic information from one cell to its daughters. This faithful transmission reflects the accuracy of DNA replication and spindle assembly, as well as mechanisms that repair damaged DNA and realign chromosomes that have been incorrectly attached to the mitotic spindle. But these processes cannot guarantee the accurate transmission of genetic information unless cells can solve the completion problem (see page 12). To do so, cells must ensure that they do not enter mitosis until DNA replication is complete and that they do not begin anaphase until the chromosomes have been correctly aligned on the spindle. Similarly, cells that have incurred DNA damage during G1 should wait until the damage has been repaired before beginning DNA replication, and cells that have suffered DNA damage during S phase or G2 should repair the damage before they enter mitosis.

Failure to solve the completion problem produces genetic damage. Cells that enter mitosis before DNA replication is complete suffer chromosome breakage or mis-segregation at anaphase (see Figure 7-7), and cells that enter anaphase with chromosomes whose sister chromatids are not attached to opposite poles produce daughters that have missing or extra chromosomes (Figure 8-1). Failure to repair DNA damage before initiating S phase or mitosis can propagate the genetic damage and lead to chromosome breakage. All of these events produce cells that are dead, mutant, or have abnormal numbers of chromosomes. In unicellular organisms, in which each cell divi-

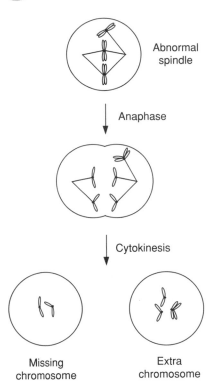

Abnormal spindle

↓ Anaphase

↓ Cytokinesis

Missing chromosome

Extra chromosome

FIGURE 8-1   Premature anaphase. Proper chromosome segregation requires that sister chromatids are attached to opposite poles of the spindle. If a cell enters anaphase before a chromosome has become attached to both poles of the spindle, both sisters will be segregated to one of the daughters.

sion reproduces the organism, genetic damage reduces the reproductive capacity of a population of cells. In multicellular animals genetic damage in the germ line reduces the production of viable eggs and sperm and gives rise to mutant progeny, while mutations in somatic cells can alter cell proliferation and lead to cancer (see Chapter 9).

To solve the completion problem, cells must coordinate the cell cycle engine with the completion of downstream events (Figure 8-2). For instance, the activation of MPF induces chromosome condensation, which in turn stops chromosome replication. Therefore, cells must ensure that the activation of MPF is delayed until DNA replication is complete. Likewise, at anaphase cells must ensure that the activation of cyclin proteolysis and the resulting inactivation of MPF does not occur until all the chromosomes are aligned correctly on the spindle.

CHECKPOINTS, FEEDBACK CONTROLS, AND RELATIVE TIMING

By 1980, experiments in yeast and frogs had produced two very different views of the completion problem. Analyzing budding yeast cdc^{ts} mutants had shown that the initiation of one cell cycle event is dependent on the completion of earlier events. In contrast, experiments performed on fertilized frog eggs had shown that a free-running engine that is insensitive to the comple-

FIGURE 8-2   Feedback controls and the cell cycle engine. The progress of the cell cycle is regulated at checkpoints that correspond to transitions in the cell cycle engine. Unreplicated DNA and damaged DNA activate feedback controls that prevent the activation of MPF and thus keep cells from entering mitosis until they have replicated their DNA. An improperly assembled spindle activates a feedback control that prevents the inactivation of MPF and thus keeps cells from going into anaphase until they have aligned their chromosomes correctly on the spindle. Finally, damaged DNA also activates a feedback control that prevents passage through Start.

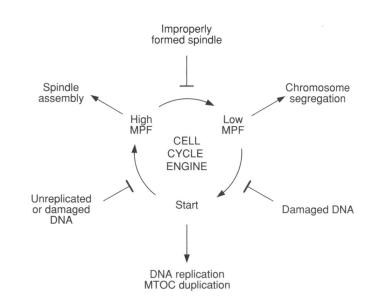

tion of downstream events drives the cell cycle. In this section we discuss how these different results reveal the two different mechanisms for solving the completion problem.

The relative timing of the cell cycle engine and downstream events solves the completion problem in frog embryos

In fertilized frog eggs, inhibiting DNA replication or spindle assembly does not arrest the cell cycle engine (see Chapter 2). Although this behavior is unusual, even in embryonic cell cycles, it reveals that the **relative timing** of the cell cycle engine and downstream events can solve the completion problem. In embryonic cell cycles, the events that lead to the activation of MPF and those that lead to DNA replication are both induced by the inactivation of MPF at the end of the preceding mitosis. As long as it takes longer for the cell cycle engine to activate MPF than it takes the downstream events to complete DNA replication, cells will not enter mitosis until they have finished replicating their DNA (Figure 8-3).

RELATIVE TIMING, DNA REPLICATION NORMAL

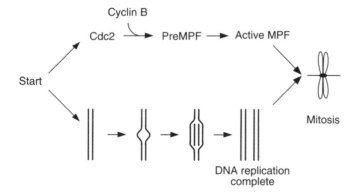

RELATIVE TIMING, DNA REPLICATION DELAYED

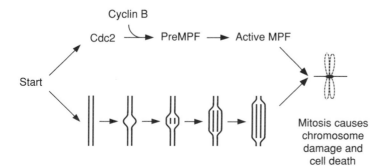

FIGURE 8-3 Solutions to the completion problem: relative timing. In frog embryos it normally takes longer to activate MPF than it does to complete DNA replication. This arrangement ensures that cells do not enter mitosis until they have completed replication. But if the completion of DNA replication is delayed or inhibited, the entry into mitosis cannot be delayed, leading to lethal damage to the chromosomes.

Relying solely on relative timing is a risky way of solving the completion problem because the window between the end of DNA replication and the beginning of mitosis is short. If DNA replication is significantly delayed, MPF will be activated before cells have finished replication, leading to genetic damage. To ensure accurate chromosome segregation, the downstream events must run faster than the cell cycle engine in every cell in an embryo. This requirement can be met only if the cell-to-cell variability in the rates of the engine and of downstream events is small. As fragile as relative timing seems, the vast majority of fertilized frog eggs develop successfully into tadpoles, demonstrating that the combination of maternal investment and evolutionary fine-tuning have produced the rigidly timed cell cycles that allow this strategy for solving the completion problem to succeed.

Feedback controls restrain the autonomous oscillation of the cell cycle engine

Even in early embryonic cell cycles relative timing is rarely the only strategy for solving the completion problem. Although blocking DNA replication cannot slow down the first few cell cycles of frog embryos, inhibiting DNA synthesis in embryos that have more than 100 nuclei arrests the cell cycle in interphase (Figure 8-4). The ability of unreplicated DNA to generate a signal that prevents the activation of MPF and entry into mitosis is an example of a feedback control arresting the cell cycle engine at a particular checkpoint (Figure 8-5). In such pathways an unfinished downstream event inhibits the ability of the cell cycle engine to pass a checkpoint that would normally lead to the next transition in the engine.

The major checkpoints at which feedback controls regulate the cell cycle are Start, entry into mitosis, and exit from mitosis. These are the same checkpoints at which the cell cycle pauses in response to signals generated outside the cell. In budding yeast, mating factors, starvation, and small cell size can all block passage through Start (see Chapters 3 and 6), and vertebrate cells cannot pass the restriction point without growth factors (see Chapter 6). Cell size and nutrients regulate entry into mitosis in fission yeast (see Chapter 4), and unfertilized frog eggs cannot exit meiosis II until they are fertilized (see Chapter 10).

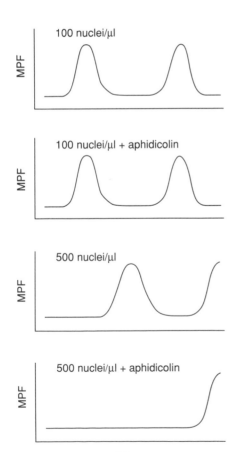

100 nuclei/μl

100 nuclei/μl + aphidicolin

500 nuclei/μl

500 nuclei/μl + aphidicolin

Time

FIGURE 8-4 Feedback controls appear during development. Experiments in frog egg extracts suggest that a feedback control that allows unreplicated DNA to block the activation of MPF appears during the first few cell cycles of fertilized frog eggs. When the density of nuclei in the extract cytoplasm is less than $100/\mu l$ ($1\mu l$ is approximately equivalent to the amount of cytoplasm in a single egg), adding aphidicolin, a DNA polymerase inhibitor, does not prevent MPF activation. With a density equivalent to 500 nuclei per egg, the cell cycle is slower because it takes longer to replicate the larger amount of DNA, and the addition of aphidicolin does block the activation of MPF and entry into mitosis. Reference: Dasso, M., & Newport, J. W. *Cell* **61**, 811–823 (1990).

FEEDBACK CONTROL, DNA REPLICATION NORMAL

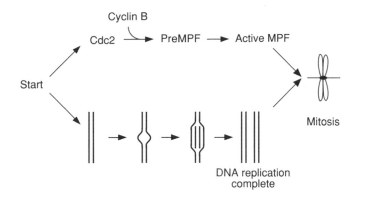

FIGURE 8-5 Solutions to the completion problem: feedback controls. Most cells have feedback controls that can arrest the progress of the cell cycle engine at particular checkpoints. In this example, a feedback control responds to the presence of incompletely replicated DNA by generating a signal that prevents the activation of MPF. This strategy allows entry into mitosis to be delayed when cells encounter problems in DNA replication.

FEEDBACK CONTROL, DNA REPLICATION DELAYED

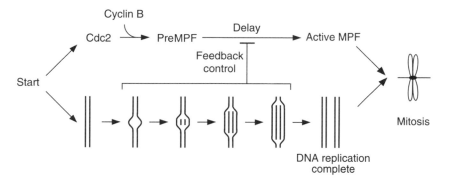

We think of the control machinery that regulates passage through checkpoints as regulatory modules that have been bolted on to an ancestral, unrestrained cell cycle engine. Each of these modules can be activated by a variety of events. Some stimuli, such as damaged or unreplicated DNA or improperly assembled spindles, arrest the vast majority of cell cycles. Others, such as the absence of sperm from unfertilized eggs or the presence of yeast mating factors, can arrest the cell cycle in only one cell type.

CONTROL OF ENTRY INTO MITOSIS

Irradiating cells with X rays or ultraviolet light produces single stranded nicks and gaps and double stranded breaks in the chromosomal DNA. If the irradiated cells are in G2, the presence of damaged DNA keeps them from entering mitosis until they have repaired the damage. The first clue that this block was due to a feedback control came from studies in mammalian cells performed in 1974. Treating irradiated tissue culture cells with caffeine abolished the damage-induced delay and allowed cells to enter mitosis before DNA repair was complete, suggesting that caffeine overcame a feedback control that was preventing cells from entering mitosis until they had repaired the DNA damage. Because the caffeine-treated cells entered mitosis before repairing their DNA, they suffered lethal chromosome damage as they passed through mitosis, providing a striking demonstration of the vital role that feedback controls play in safeguarding the integrity of genetic information.

Studies of radiation-sensitive yeast mutants identify a feedback control gene

Because caffeine inhibits many biochemical reactions, determining how it overcomes the feedback control that detects damaged DNA is difficult. A more profitable approach to understanding this feedback control involves analysis of radiation-sensitive *(rad)* mutants in budding yeast. The *rad*⁻ mutants were isolated as strains that were unusually sensitive to irradiation and were originally presumed to identify components in the enzymatic machinery that repairs damaged DNA. The effects of caffeine on mammalian cells suggested an alternative possibility: that some of the *rad*⁻ mutants were defective in a feedback control that detected damaged DNA, rather than in the enzymes that repair DNA damage. These two possibilities were distinguished by determining the responses of different *rad*⁻ mutants to irradiation (Figure 8-6). Irradiated wild-type cells arrested in G2 until they had repaired any DNA damage and then proceeded through mitosis to yield viable cells. Cells that could not repair DNA damage arrested permanently in G2. Strikingly, cells of one mutant, *rad9*⁻, went through mitosis without waiting to repair their broken chromosomes and died as a result, suggesting that *rad9*⁻ is a feedback control mutant. Further experiments confirmed that in *rad9*⁻ cells, damaged DNA can be repaired but cannot prevent passage through mitosis.

Detailed genetic analysis of the feedback control pathway that detects DNA damage has led to a number of important conclusions. First, the feedback control is exquisitely sensitive. A single double-stranded DNA break is all it takes to keep wild-type budding yeast cells from entering mitosis. Second, many of the *cdc*ᵗˢ mutants that arrest the cell cycle do so by activating feedback controls. For example, the *cdc9*ᵗˢ mutant is a temperature-sensitive mutation in DNA ligase, the enzyme that joins single-stranded stretches of replicated DNA during S phase. The *cdc9*ᵗˢ mutant arrests in G2 and remains viable at the restrictive temperature, but the *cdc9*ᵗˢ *rad9*⁻ double mutant behaves quite differently. The double mutant cells divide on schedule at the restrictive temperature and die as they do so, demonstrating that the lack of DNA ligase arrests the cell cycle by activating a feedback control pathway. Third, cells can distinguish between different aspects of DNA metabolism. Although *rad9*⁻ mutants cannot respond to DNA damage, they still arrest in response to unreplicated DNA.

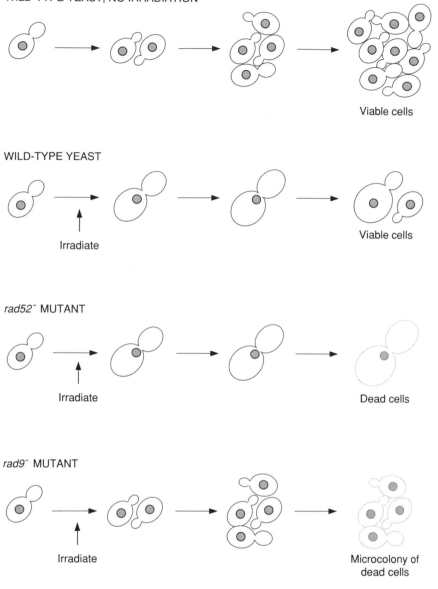

WILD-TYPE YEAST, NO IRRADIATION

Viable cells

WILD-TYPE YEAST

Irradiate

Viable cells

rad52⁻ MUTANT

Irradiate

Dead cells

rad9⁻ MUTANT

Irradiate

Microcolony of
dead cells

FIGURE 8-6 Assay for feedback control mutants. The response of individual budding yeast cells to irradiation distinguishes two types of radiation-sensitive *(rad)* mutants from wild-type cells. Wild-type cells that are irradiated with X rays in G2 show a delay in cell division that reflects the time taken to repair damaged DNA. Mutations such as *rad52⁻* that cannot repair DNA damage arrest permanently in response to the feedback control that detects DNA damage. Mutations such as *rad9⁻* that inactivate the feedback control that detects DNA damage do not delay before dividing and therefore suffer lethal chromosome damage. Reference: Weinert, T. A., & Hartwell, L. H. *Science* **241**, 317–322 (1988).

The pathway that arrests the cell cycle in response to irradiated DNA is complex. In principle, feedback controls must consist of at least three components: a sensor that monitors the completion of the downstream event, a signal that this sensor generates, and a response element in the cell cycle engine that causes it to arrest or delay at a checkpoint. In practice, however, feedback controls have many more components: genetic screens have identified five genes other than *rad9⁺* whose products participate in feedback control that detects damaged DNA in budding yeast. Mutants in three of these genes behave like *rad9⁻*: They cannot respond to DNA damage but arrest normally in response to unreplicated DNA. Two other mutants, *mec1⁻* and *mec2⁻*, fail to arrest in response to either DNA damage or unreplicated DNA

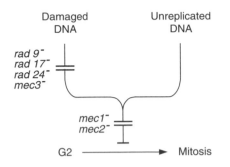

FIGURE 8-7 Budding yeast response to damaged and unreplicated DNA. The known mutations that allow budding yeast cells with damaged DNA to enter mitosis are shown. Four of these mutants (*rad9⁻*, *rad17⁻*, *rad24⁻*, and *mec3⁻*) still prevent cells with unreplicated DNA from going through mitosis, but two others (*mec1⁻*, and *mec2⁻*) can no longer respond to unreplicated DNA. The step in the cell cycle engine that these mutants regulate is not known. Reference: Weinert, T. A. *Radiation Res.* **132**, 141–143 (1992).

(Figure 8-7). This analysis suggests that although different sensors detect damaged and unreplicated DNA, these two feedback control pathways converge to regulate a common step in the progress of the cell cycle engine. The molecular basis of the feedback control that detects damaged DNA remains to be elucidated. How does the cell sense the presence of unreplicated or damaged DNA, what intracellular signals do these lesions generate, and how do these signals arrest the cell cycle engine? Although the *RAD9⁺* gene has been cloned and sequenced, its sequence does not reveal any similarity to known genes that would hint at how it participates in arresting the cell cycle.

A feedback control keeps G1 cells with damaged DNA from entering S phase, and this control differs from the one that prevents entry into mitosis. In budding yeast, the *rad9⁻* mutation allows G2 cells with damaged DNA to enter mitosis, but G1 cells with damaged DNA cannot enter S phase. Mutants in the mammalian *p53* gene have exactly the opposite phenotype: Mutant G1 cells with damaged DNA can initiate DNA replication, but DNA damage in G2 cells keeps them from entering mitosis (see pages 162–164).

Feedback controls and relative timing are redundant solutions to the completion problem

Is the feedback control that *rad9⁻* mutants identify essential for successful progress through each cell cycle? Cells that lack the *RAD9⁺* gene grow and divide normally, showing that the feedback control that detects damaged DNA is not needed to produce viable cells. Because radiation-induced DNA damage is a rare event, this finding does not seem surprising. But in every cell cycle, DNA replication produces transient structures that the cell recognizes as damaged DNA, such as the single-stranded nicks and gaps in newly replicated DNA. These lesions, like those induced by radiation, must be repaired before every mitosis. Thus, the ability of Δ*rad9* cells to survive in spite of the DNA damage that inevitably accompanies DNA replication suggests that feedback controls and relative timing are redundant solutions to the completion problem. Even in the absence of the feedback control, the faster pace of downstream events compared with that of the cell cycle engine ensures that most cells do not pass through mitosis until they have repaired the replication-induced DNA damage.

The presence of two mechanisms for solving the completion problem allows each mechanism to protect the cell from failures in the other. If the unrestrained cell cycle engine ran faster than the downstream events, feedback controls would have to slow the engine in every cell cycle, and occasional failures in the controls would lead to genetic damage. If cells relied on relative timing alone, random variations in the relative rates of the cell cycle engine and downstream events would damage the minority of cells in which the downstream events ran slower than the cell cycle engine. Mutations that eliminate feedback controls have precisely this effect: The cells remain viable, but genetic information is much less accurately propagated. For example, even in the absence of radiation, Δ*rad9* mutants lose chromosomes 20 times more frequently than wild-type cells.

Mutations that accelerate the cell cycle engine affect its response to damaged or unreplicated DNA

Mutations that allow fission yeast to enter mitosis with damaged or unreplicated DNA fall into two classes (Figure 8-8). The first class consists of mutations, like the budding yeast *mec1⁻* mutation, that probably alter components of the signaling pathway. These mutations do not reduce the threshold size at which cells normally enter mitosis, but they block the ability of damaged or unreplicated DNA to delay entry into mitosis. All but one of the known fission yeast mutants of this type are insensitive to both unreplicated and damaged DNA (Figure 8-9).

The second class consists of mutations that reduce the ability of the cell cycle engine to respond to signals generated by feedback controls. The most striking examples are mutations in fission yeast that reduce the tyrosine phosphorylation of Cdc2 that plays a critical role in preventing the premature activation of MPF (see Chapter 4). These mutations reduce the threshold size

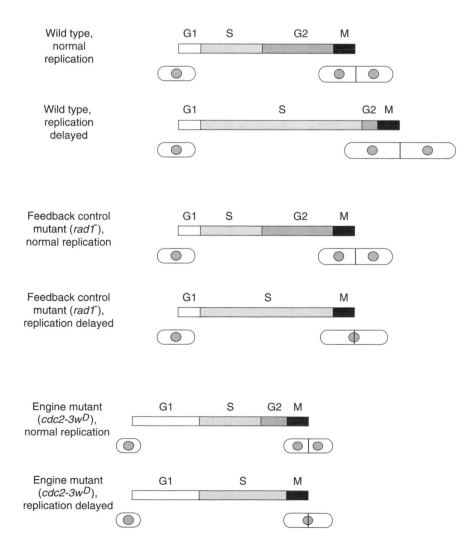

FIGURE 8-8 Different types of feedback control mutant. The responses of three different fission yeast strains to reducing the rate of DNA replication are shown. Wild-type cells delay mitosis until the DNA has been replicated. The *rad1⁻* mutation is an example of a mutation that prevents cells from generating a signal that they contain unreplicated DNA: The mutation does not alter the threshold size at which cells whose DNA is intact enter mitosis, but *rad1⁻* strains cannot delay mitosis in response to unreplicated DNA. The *cdc2-3w^D* mutation is an example of a mutation that keeps the cell cycle engine from responding to signals generated by unreplicated DNA: The mutation reduces the threshold size for mitosis, and *cdc2-3w^D* strains cannot delay mitosis in response to unreplicated DNA. References: Rowley, R., Subramani, S., & Young, P. G. *EMBO J.* **11**, 1335–1342 (1992); Enoch, T., Carr, A., & Nurse, P. *Genes and Dev.* **6**, 2035–2046 (1992).

FIGURE 8-9 Fission yeast response to damaged and unreplicated DNA. Many mutants that affect the response of fission yeast to damaged or unreplicated DNA have been isolated. The mutants may be divided into two classes. Members of one class abolish the ability of unreplicated or damaged DNA to delay entry into mitosis but have no effect on the normal cell cycle. The other class is mutations (boxed) that accelerate the activation of MPF; all of these reduce the threshold size for entering mitosis in untreated cells. References: Al-Khodairy, F., & Carr, A.M. *EMBO J.* **11**, 1343–1350 (1992); Rowley, R., Hudson, J., & Young, P. G. *Nature* **356**, 353–355 (1992); Enoch, T., Carr, A. M., & Nurse, P. *Genes and Dev.* **6**, 2035–2046 (1992).

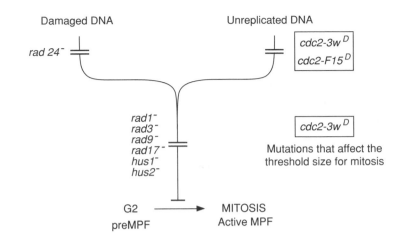

for entry into mitosis, and although unreplicated DNA may delay mitosis briefly, the delay is too short for cells to recover. One such mutation is $cdc2$-$3w^D$, which increases the activity of Cdc2 and allows cells to enter mitosis in the absence of Cdc25. Cells that harbor this mutation enter mitosis at a smaller size than wild-type cells and cannot delay entry into mitosis in response to damaged DNA.

The ability of changes in tyrosine phosphorylation of Cdc2/28 to affect the response to unreplicated DNA is not universal. Mutations that prevent tyrosine phosphorylation of budding yeast Cdc28 have no effect on the ability of unreplicated DNA to arrest the cell cycle. We cannot yet explain this difference but speculate that it means that unreplicated DNA does not directly regulate the enzymes that control tyrosine phosphorylation of Cdc2. Instead, we believe that unreplicated DNA activates some other pathway that inhibits entry into mitosis. This control system can be overridden by dephosphorylation of Cdc2/28 in fission yeast, but not in budding yeast. Understanding exactly how DNA replication controls the activation of MPF and entry into mitosis remains an important area for future research.

Cells that continue to grow while arrested by feedback controls induce a program of cell death

In some mammalian cells feedback controls appear to regulate cell growth in addition to the cell cycle engine. When synchronous populations of human cells are treated with DNA synthesis inhibitors during S phase, the cells cease growing and the rate of protein synthesis decreases at about the same time that control cells enter G2. The replication-blocked cells do not suffer lethal damage and begin to grow and proliferate again when DNA synthesis is allowed to resume. This result suggests that in human cells inhibitors of DNA replication arrest cell growth as well as the progress of the cell cycle engine.

A very different response is seen in cultured rodent cells, which continue to grow indefinitely after DNA synthesis has been blocked, eventually reaching sizes much greater than those of normal cells. If the block to DNA synthesis is prolonged, cells die in one of two ways when the block is released. About half the cells die without ever undergoing mitosis, and the other half undergo aberrant mitoses and die shortly afterward, probably as a consequence of massive genetic damage (Figure 8-10). But if protein synthesis is blocked while DNA replication is inhibited, the cells do not grow and do not die when replication resumes. This response suggests that death of the cells in which protein synthesis was normal is connected in some way with their large size. One explanation for why large cells perform aberrant mitoses is that as cells get bigger, the ability of a feedback control to hold them at checkpoints declines. In this scenario the amount of unreplicated DNA would decline as replication resumed, weakening the negative signal generated by the feedback control. As replication continued, the amount of unreplicated DNA would fall to a level at which the feedback control could no longer oppose the influence of cell size, and the cell would enter mitosis even though DNA replication was incomplete. Yeast cells that have been arrested by a single, irreparable double-stranded break in their DNA behave similarly. The cells remain arrested in G2 for several hours while they continue to grow, but eventually they enter mitosis, divide, and die.

The rodent cells that respond to being released from the replication block by dying without going through mitosis may be inducing their own death. In multicellular organisms such cell death is an adaptive response that kills cells that are liable to produce genetically damaged progeny, such as those that have reached a size where feedback controls can no longer reliably arrest them. Although individual cells die, the organism avoids producing mutant cells that could give rise to cancer. Such deliberately induced cell death, called **apoptosis,** occurs in many situations. Although the exact mechanism of apoptosis is not well understood, genes that are required for programmed cell death and other genes that can prevent death have been identified. Lymphocytes treated with microtubule polymerization inhibitors do not show prolonged arrest in mitosis but undergo apoptosis instead of continuing in the cell cycle. When treated with the same inhibitors, cells that overexpress the $bcl2^+$ (B cell lymphoma) gene, which prevents apoptosis, do not die. Expression of $bcl2^+$ can also prevent apoptosis caused by expressing c-myc^+ in serum-starved cells. The ability of genes like $bcl2^+$ to rescue cells destined for death probably plays a crucial role in the events that give rise to cancer (see page 157).

Apoptosis also plays important roles in development. In many organisms substantial cell death occurs during development. In the nematode *Caenorhabditis elegans,* which has rigidly determined cell lineages, the cells that die are the same in different individuals, showing that these particular cells are predestined to die. Apoptosis is especially prominent in tissue remodeling, formation of the nervous system, and modulation of the immune response. It is not yet certain, however, that a single mechanism can explain all the different situations in which cells are induced to die.

FIGURE 8-10 Induced cell death. The responses of cultured human and rodent cells to inhibition of DNA synthesis differ. Human cells cease growing when DNA synthesis is inhibited but resume growth and DNA synthesis to produce viable progeny when the block to replication is lifted. By contrast, when rodent cell lines are treated with DNA synthesis inhibitors the cells do not stop growing. When the block to replication is lifted, the cells respond in one of two ways: About half of the cells die without passing through mitosis as a result of induced cell death (apoptosis), and the remainder undergo an aberrant mitosis and produce two dead progeny. Using an inhibitor of protein synthesis prevents the rodent cells from growing and allows them to survive a prolonged period of S phase arrest. Reference: Schimke, R. T., Kung, A. L., Rush, D. F., & Sherwood, S. W. *Cold Spring Harbor Symp. Quant. Biol.* **66**, 417–425 (1991).

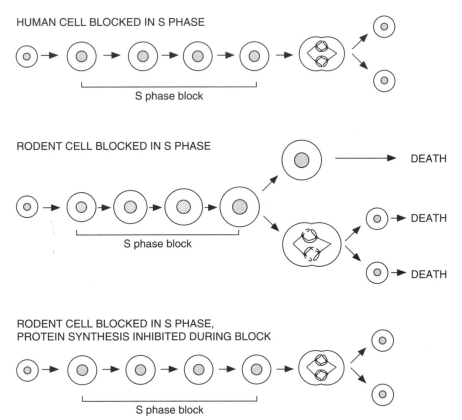

CONTROL OF EXIT FROM MITOSIS

In most cells, treatment with microtubule-depolymerizing drugs induces a mitotic arrest by activating a feedback control that detects improperly assembled spindles. A number of budding yeast mutants, called *mad⁻* (mitotic arrest deficient) or *bub⁻* (budding uninhibited by benzimadazole), destroy the feedback control. Treating these mutants with microtubule-depolymerizing drugs (such as benomyl) does not prevent the inactivation of MPF and exit from mitosis and leads to massive genetic damage that kills the cells (Figure 8-11).

Kinetochores that are not correctly attached to microtubules can arrest cells in mitosis

What defect does the feedback control that monitors the spindle detect? Mutations that increase or decrease microtubule stability, inactivate microtubule motors, or inhibit the function of the spindle pole body all cause cell

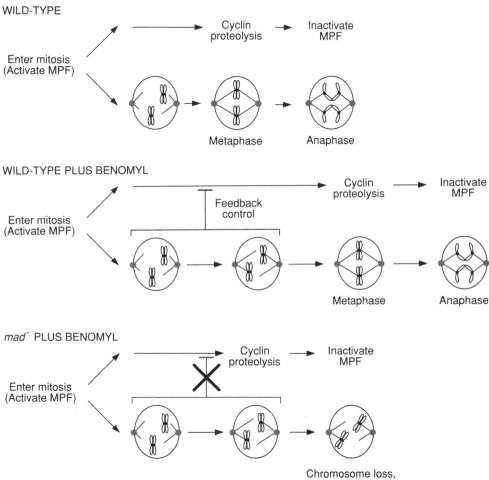

FIGURE 8-11 Mitotic-arrest-deficient mutants. Activating MPF initiates assembly of the spindle and alignment of the mitotic chromosomes at the metaphase plate. In wild-type budding yeast cells treated with low doses of benomyl (a microtubule polymerization inhibitor), the activation of the cyclin proteolysis machinery is delayed in response to a feedback control that monitors spindle assembly. This control is inactivated in *mad*⁻ and *bub*⁻ strains, but the mutants can grow in the absence of microtubule inhibitors because the events that lead to the inactivation of MPF normally take longer than spindle assembly. In the presence of benomyl the mutants cannot delay the activation of the cyclin proteolysis machinery, and anaphase occurs before the chromosomes are properly aligned, producing dead cells. References: Hoyt, M. A., Trotis, L., & Roberts, B. T. *Cell* **66**, 507–517 (1991); Li, R., & Murray, A. W. *Cell* **66**, 519–531 (1991).

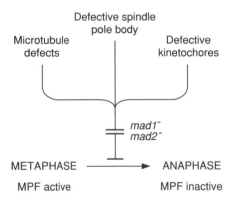

FIGURE 8-12 Spindle defects activate mitotic feedback controls. The *mad1⁻* and *mad2⁻* mutants inactivate a feedback control that detects several different types of errors in spindle assembly. These include aberrant numbers of microtubules, defective spindle pole bodies, and defective kinetochores. The molecular basis of this feedback control is not understood. Reference: Li, R., & Murray, A. W. unpublished.

cycle arrests that are abolished by *mad⁻* mutants (Figure 8-12). One explanation of how these different lesions are detected by the same feedback control is that they all affect a particular aspect of the spindle. An especially attractive candidate is the interaction between kinetochores and microtubules.

Circumstantial evidence suggests that the budding yeast does indeed monitor the behavior of kinetochores. Small circular plasmids that carry centromeres, replication origins, and selectable genes have been constructed. These minichromosomes partially mimic the behavior of real chromosomes: Real chromosomes and minchromosomes are both present in one copy per cell, but errors in minichromosome segregation occur once in every 100 cell divisions, compared with once in every 100,000 divisions for a real chromosome. A genetic trick can increase the number of the minichromosomes from one to four per cell. The presence of these extra minichromosomes profoundly delays cells in mitosis, and this delay is abolished by *mad⁻* mutants, demonstrating that activation of the feedback control induces the delay (Figure 8-13).

Why do extra copies of minichromosomes delay cells in mitosis? One explanation is that the mere presence of extra centromeres in the cell is enough to alter spindle function. Another possibility is that minichromo-

FIGURE 8-13 Minichromosomes delay anaphase. Small circular plasmids, or minichromosomes (about 10,000 base pairs), partially mimic the behavior of natural chromosomes. When there is only one copy per cell, minichromosomes do not delay anaphase, but when the number of copies is increased to four per cell, completion of mitosis is profoundly delayed. This delay is abolished in *mad1⁻* and *mad2⁻* mutants, showing that it results from activation of a feedback control. Reference: Wells, W., & Murray, A. W., unpublished observations.

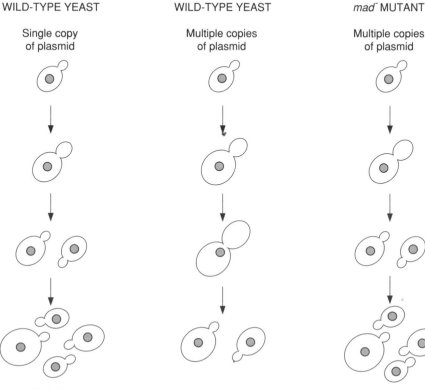

somes delay cells in mitosis because they are partially defective in some aspect of normal chromosome behavior. We favor the second explanation because in other cells, single chromosomes that occupy abnormal positions on the spindle can prevent anaphase. One example is the behavior of the cells that produce sperm in praying mantids. During meiosis I, the three sex chromosomes are normally linked to each other and lie on the metaphase plate. In about 10% of meiotic cells, the linkage between the sex chromosomes decays prematurely, and one of them leaves the metaphase plate and takes up a position close to one pole of the spindle. Even though the kinetochore of this chromosome is still attached to microtubules, the cells never enter anaphase, suggesting that a functional kinetochore that lies close to the spindle pole activates a feedback control that prevents the onset of anaphase (Figure 8-14). In yeast cells the small size of minichromosomes may allow them to take up positions close to the spindle poles that activate such a feedback control.

The behavior of mutants in γ tubulin also suggests that the feedback control over the exit from mitosis involves the spindle poles. The microtubule organizing centers contain γ tubulin, which appears to play a crucial role in nucleating the assembly of microtubules (see page 75). When *Aspergillus* spores that lack the γ-tubulin gene germinate, they fail to assemble microtubules, but they enter and leave mitosis even though no spindle is visible. Since wild-type *Aspergillus* spores can be successfully arrested in mitosis by micro-

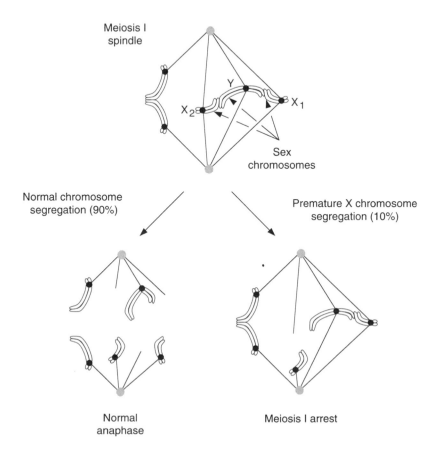

FIGURE 8-14 A meiotic feedback control. Praying mantids have three sex chromosomes, a Y chromosome and two X chromosomes. In meiosis I in males, the two X chromosomes recombine with the Y chromosome and form a group of three chromosomes on the metaphase plate. In about 10% of meioses, one of the X chromosomes becomes detached from the Y chromosome and takes up a position close to one of the spindle poles. These cells remain arrested in metaphase indefinitely, demonstrating that they have a feedback control that can monitor the positions of chromsomes on the spindle. Reference: Nicklas, R. B., & Arana, P. *J. Cell Sci.* **102**, 681–690 (1992).

FIGURE 8-15 Mitotic feedback control and γ tubulin. Wild-type *Aspergillus* spores germinate to form multinucleate cells. Cells that germinate in the presence of benomyl lack microtubules and arrest at mitosis. γ tubulin is an essential component of the spindle pole body that probably nucleates microtubules. When *Aspergillus* spores that lack γ tubulin germinate, they enter mitosis and fail to form a mitotic spindle. Despite this failure, the cells are not arrested in mitosis, demonstrating that the spindle pole body plays an essential role in the feedback control that monitors spindle assembly. Reference: Oakley, B. R., Oakley, C. E., Yoon, Y., & Jung, M. K. *Cell* **61**, 1289–1301 (1990).

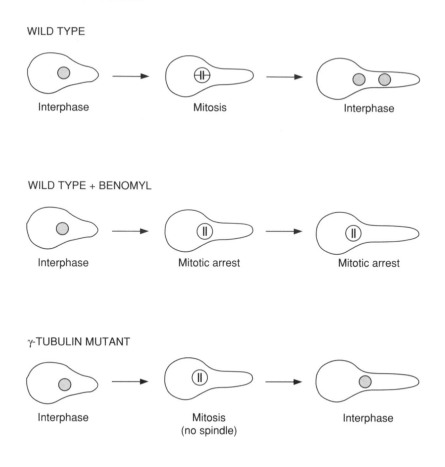

tubule depolymerizing drugs, the phenotype of the γ-tubulin mutation suggests that γ tubulin at the spindle pole body plays a critical role in allowing the cell to monitor spindle assembly (Figure 8-15).

Two members of the mitotic feedback control pathway are enzymes that modify proteins

A considerable amount of molecular analysis has been performed on the *mad⁻* and *bub⁻* mutants. All three *BUB1⁺* genes have been sequenced. *BUB1⁺* shows homology to protein kinases, and *BUB2⁺* and *BUB3⁺* show no detectable homology to other proteins. The substrates that Bub1 phosphorylates have not been identified.

Mad2 is the α subunit of a member of the group of enzymes called **protein isoprenyl transferases**. These enzymes are α–β heterodimers that add C_{15} or C_{20} carbon chains, called **prenyl groups** (more accurately polyisoprenoid groups), near the carboxyl termini of a number of proteins (Fig 8-16). These hydrophobic carbon chains resemble the side chains of membrane lipids and insert into the membranes to anchor the prenylated proteins at the surface of the membranes. Prenylated proteins include nuclear lamins, a number of small G proteins, including Ras, and the γ subunits of trimeric G proteins. The

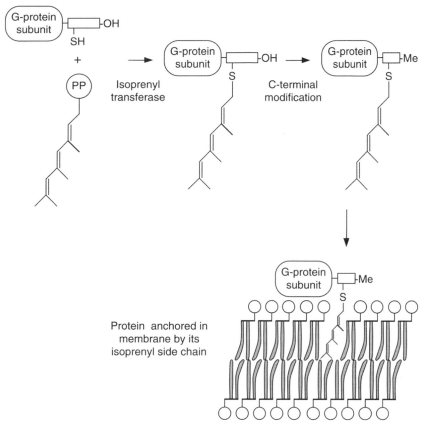

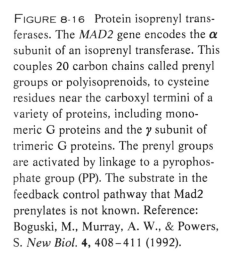

FIGURE 8-16 Protein isoprenyl transferases. The *MAD2* gene encodes the **α** subunit of an isoprenyl transferase. This couples 20 carbon chains called prenyl groups or polyisoprenoids, to cysteine residues near the carboxyl termini of a variety of proteins, including monomeric G proteins and the **γ** subunit of trimeric G proteins. The prenyl groups are activated by linkage to a pyrophosphate group (PP). The substrate in the feedback control pathway that Mad2 prenylates is not known. Reference: Boguski, M., Murray, A. W., & Powers, S. *New Biol.* **4,** 408–411 (1992).

inability of *mad2⁻* mutants to arrest in mitosis reveals a role for prenylated proteins in the feedback control over exit from mitosis. The substrates whose prenylation is required for maintaining the integrity of the feedback control have not been identified.

CONCLUSION

Analyzing how different cells solve the completion problem has helped to explain the very different appearance of early embryonic and somatic cell cycles. Frog eggs solve the completion problem by relying solely on a cell cycle engine that runs slower than the downstream events that it controls. As a result, inhibiting downstream events has no effect on the progress of the cell cycle engine. In addition to relying on the relative timing of the cell cycle engine and downstream events, somatic cell cycles use feedback controls that arrest the engine at checkpoints until downstream events have been completed. The molecular analysis of feedback controls is still in its infancy, so we know very little about how cells monitor DNA replication, DNA damage, or the structure of the mitotic spindle. The signals that these monitoring mechanisms generate and how they arrest the cell cycle engine are also exciting areas for future investigation. Although at the molecular level our understanding of feedback controls is still weak, in Chapter 9 we shall see that lesions in these controls have already been shown to play an important role in the genesis and progression of cancer.

SELECTED READINGS

Reviews

Hartwell, L. H., & Weinert, T. A. Checkpoints: controls that ensure the order of cell cycle events. *Science* **246**, 629–634 (1989).

Murray, A. W. Creative blocks: cell cycle checkpoints and feedback controls. *Nature* **359**, 599–604 (1992). *Two complementary reviews that discuss how checkpoints and feedback controls are used to regulate progress of the cell cycle engine.*

Raff, M. C. Social controls on cell survival and cell death. *Nature* **356**, 397–400 (1992).

Ellis, R. E., Yuan, J. Y., & Horvitz, H. R. Mechanisms and functions of cell death. *Ann. Rev. Cell Biol.* **7**, 663–698 (1991). *Two reviews that discuss the circumstances under which cells are programmed to die and the mechanism by which they do so.*

Original Articles

Lau, C. C., & Pardee, A. B. Mechanism by which caffeine potentiates lethality of nitrogen mustard. *Proc. Natl. Acad. Sci. USA* **79**, 2942–2946 (1982). *A demonstration that caffeine overcomes the feedback control by which damaged DNA prevents entry into mitosis.*

Weinert, T. A., & Hartwell, L. H. The *RAD9* gene controls the cell cycle response to DNA damage in Saccharomyces cerevisiae. *Science* **241**, 317–322 (1988). *The discovery that the budding yeast rad9⁻ mutant is a feedback control mutant.*

Enoch, T., & Nurse, P. Mutation of fission yeast cell cycle control genes abolishes dependence of mitosis on DNA replication. *Cell* **60**, 665–673 (1990). *Mutations that alter the regulation of the cell cycle engine can destroy feedback controls.*

Schimke, R. T., Kung, A. L., Rush, D. F., & Sherwood, S. W. Differences in mitotic control among mammalian cells. *Cold Spring Harbor Symp. Quant. Biol.* **66**, 417–425 (1991). *Comparative analysis of the response of human and rodent cells to inhibitors of DNA synthesis.*

Li, R., & Murray, A. W. Feedback control of mitosis in budding yeast. *Cell* **66**, 519–531 (1991). *The isolation of mutants that inactivate the feedback control over the exit from mitosis.*

CANCER

CANCER IS a complex and frightening disease that kills one in five adults in developed countries. Studies on the cell cycle and investigations of cancer have been intertwined for many years. Since the beginning of this century, cancer research has greatly increased our understanding of cell growth and proliferation. More recently, advances in our understanding of the cell cycle have made major contributions to elucidating how the genetic changes in cancer arise and how these changes alter the regulation of the cell cycle engine. This chapter describes how cancer research identified proteins that regulate cell proliferation, how the genetic changes that alter the expression or structure of these proteins arise, and what the prospects are that our new knowledge will lead to improvements in the diagnosis and treatment of cancer.

During a person's life there are about 10^{16} cell divisions in their body. If cancer were induced by a single genetic change that led to uncontrolled cell proliferation, this enormous number of cell divisions would produce thousands of different cancers in each individual. In reality, however, cancer is relatively rare considering the number of cells at risk. The incidence of cancer in humans increases very steeply with increasing age, and statistical analysis of the increase suggested that the development of malignant cancer requires about five independent events (Figure 9-1). The conclusion that several genetic changes must occur to produce malignant cells helped to explain why

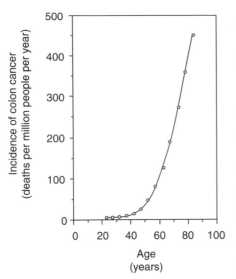

FIGURE 9-1 Cancer incidence increases with age. The incidence of cancer increases as the fifth power of age, suggesting that five independent mutational events are required to produce malignant cancer. Reference: Cairns, J. *Cancer: Science and Society* (W. H. Freeman, New York, 1978).

cells so rarely become cancerous and raised a number of important questions. What genes are altered in cancer cells, how do these changes produce cancer, and how similar are the genetic events in different cases of the same type of cancer?

ONCOGENES AND TUMOR SUPRESSOR GENES

The genetic changes that give rise to cancer affect several different properties of cells. This section discusses the genetic changes that increase cell proliferation by decreasing the dependence of cells on external growth factors, preventing senescence, or inhibiting programmed cell death. Although these changes lead to the abnormal proliferation that characterizes transformed cells, the tumors that such cells form remain benign until further genetic alterations occur. One type of alteration allows tumor cells to avoid destruction by components of the immune system that recognize and kill abnormal cells. A second type enables solid tumors to produce factors that induce the formation of new blood vessels that invade the tumor, allowing it to grow. Until this change occurs, the tumor remains benign and cannot grow beyond a certain point because the cells in its center are starved for nutrients. Finally, cells in solid tumors can acquire the ability to invade blood vessels, migrate through the body, and settle in new locations to give rise to secondary tumors.

The mutations that result in unrestrained cell proliferation occur in two classes of genes: proto-onocogenes and tumor supressor genes (see Chapter 6). In normal cells the products of proto-oncogenes act at many points along the pathways that stimulate cell proliferation, whereas tumor supressor genes encode proteins that inhibit cell proliferation and ensure the stability of the genome. Tumor cells free themselves of the normal controls on proliferation by accumulating genetic changes that activate proto-oncogenes and inactivate tumor supressor genes. The changes that activate proto-oncogenes are of two types, mutations that produce an altered protein product with increased activity and events that dramatically increase the expression of an unaltered proto-oncogene. The enormous complexity of the mammalian genome and the many types of cancer made the identification of genetic changes in cancer a seemingly impossible task, but a number of shortcuts have led to surprisingly rapid progress in isolating genes involved in cancer. The first breakthrough—the discovery that cancer could be virally transmitted—suggested that some viruses carried oncogenes, greatly simplifying the analysis of these important genes.

RNA tumor viruses carry oncogenes that induce cell proliferation

In 1910 Francis Peyton Rous identified a virus, called Rous sarcoma virus (RSV), that causes cancer in chickens. The same virus was later shown to infect and transform chicken tissue culture cells in vitro (Figure 9-2). A related virus could infect and replicate in chicken cells without transforming

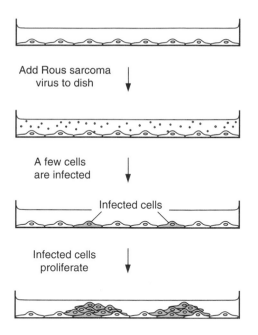

them, suggesting that Rous sarcoma virus carries an oncogene that the benign virus lacks. In 1976 advances in molecular biology made it possible to compare the genomes of these two viruses (Figure 9-3), and the transforming virus was found to contain a single extra gene, which was named *v-src* (viral-sarcoma). Introducing the *v-src* gene into established tissue culture cell lines transformed them, showing that *v-src* is an oncogene.

How had Rous sarcoma virus acquired a gene that could transform cells? Careful analysis revealed that the genome of normal cells contains a gene called *c-src* (cellular src) that is closely related to *v-src*. The cellular gene is a proto-oncogene that encodes a tyrosine kinase that does not interfere with normal cell growth. The appearance of *v-src* is believed to be the result of an event in which a precursor to Rous sarcoma virus acquired a copy of the *c-src* gene. Through many generations this ancestral virus acquired multiple mutations in the *src* gene that deregulated its tyrosine kinase activity and converted it into the oncogene *v-src*. The activated viral gene is dominant to its normal cellular counterpart.

ROUS-ASSOCIATED VIRUS (RAV)

Structural protein (*gag*)	Reverse transcriptase (*pol*)	Structural protein (*env*)

ROUS SARCOMA VIRUS (RSV)

Structural protein (*gag*)	Reverse transcriptase (*pol*)	Structural protein (*env*)	Transforming gene (*src*)

FIGURE 9-3 Structure of Rous sarcoma virus. The genetic structure of Rous sarcoma virus (RSV) and the nontransforming Rous-associated virus (RAV) are shown. The genomes of both viruses consist of a single-stranded RNA molecule that is copied into DNA inside the host cell. Rous sarcoma virus contains an extra gene called *v-src* that is responsible for its ability to transform cells. Reference: Stehelin, D., Varmus, H. E., Bishop, J. M., & Vogt, P. K. *Nature* **260**, 170–173 (1976).

Rous sarcoma virus is a **retrovirus.** Retroviruses use RNA as their genome but copy it into DNA when they infect cells. Many members of this family are RNA tumor viruses that carry oncogenes that have arisen from a cellular proto-oncogene that the virus has captured and mutated during evolution. Since most retroviruses replicate without killing their hosts, the presence of oncogenes may give RNA tumor viruses a selective advantage by inducing the cells that they infect to proliferate.

Some DNA viruses also carry oncogenes, but these genes are not obviously related to cellular genes. In viruses like SV40 and adenoviruses (which cause coldlike symptoms) the physiological role of the oncogenes is to induce the cell to enter S phase so that the host cell's replication machinery can replicate the virus. Since viral replication leads to cell death, these genes have not been selected to stimulate proliferation of the host cells and act as oncogenes only when introduced into cells in which viral replication is blocked.

DNA from tumors that contain oncogenes can transform normal cells

Although studies on retroviruses have identified many oncogenes, most human tumors are not induced by viral oncogenes. Nevertheless, the ability of viral oncogenes to transform cells suggested that dominant mutations might convert proto-oncogenes into oncogenes that would initiate the progression to cancer. To test this hypothesis, chromosomal DNA isolated from tumors was transfected into normal tissue culture cells. Most of the recipient cells received pieces of DNA that had no effect on their growth. A tiny fraction, however, acquired a segment of DNA carrying the dominant oncogene, became transformed and lost contact inhibition. As a result, these cells

FIGURE 9-4 Finding oncogenes in transformed cells. To assay whether tumor cells contain active oncogenes, their DNA is extracted and used to transfect normal cells. If some of the transfected cells become transformed, they have acquired a dominant oncogene. By tagging the transfected DNA it is ultimately possible to recover and characterize the oncogenes that these experiments identify. Reference: Shilo, B. Z., & Weinberg, R. A. *Nature* **289,** 607–609 (1981).

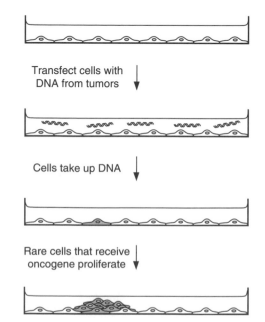

Transfect cells with
DNA from tumors

Cells take up DNA

Rare cells that receive
oncogene proliferate

formed small foci of proliferating cells in a lawn of contact-inhibited cells (Figure 9-4). Several clever genetic tricks make it possible to isolate oncogenes identified in this way, although this task is still much harder than isolating viral oncogenes.

The first oncogenes identified in tumor DNA were dominant mutations in the *c-ras* gene whose viral counterpart had already been identified by studies on retroviruses. Mutations in *c-ras* are among the most common found in human tumors, and most of them inactivate the GTPase activity of the Ras protein. As a result the majority of the protein is in the GTP-bound form, even when growth factors are absent, stimulating the signaling pathways that drive cells through the restriction point (see pages 109–111).

The ability of oncogenes to transform cells suggests that the proto-oncogenes from which they are derived help to regulate cell proliferation. Molecular analysis has spectacularly confirmed this prediction. Proto-oncogenes encode proteins involved in all aspects of controlling cell proliferation: growth factors, growth factor receptors, proteins that generate second messengers, protein kinases, and early and delayed response genes.

Common chromosomal abnormalities in tumors can identify proto-oncogenes

During the formation of B cells, genetic rearrangements assemble antibody genes from several segments of DNA. Some of these events are abnormal: They join a proto-oncogene on one chromosome to a segment of an antibody gene on another. Because these **translocations** link a proto-oncogene to a segment of DNA that strongly stimulates transcription, they greatly increase the amount of proto-oncogene product in the cell. Although these events are extremely rare, if overexpression of the proto-oncogene leads to uncontrolled cell proliferation, the cells that have undergone the translocation will multiply and eventually give rise to a tumor. Certain chromosomal translocations are found in many different tumors of white blood cells (leukemias and lymphomas), suggesting that proto-oncogenes lie at the point at which a chromosome has become joined to part of an antibody gene.

Two important proto-oncogenes have been identified by studying chromosomal translocations. One is the cyclin D1 gene which is activated in a large fraction of one form of lymphoma. This finding offers strong evidence that overexpression of this G1 cyclin can alter the regulation of mammalian cell proliferation in much the same way that overexpression of G1 cyclins in yeast alters the control of Start. In a different form of lymphoma, a translocation that activates the *bcl2* gene is found in 90% of all cases. We have seen that expression of this gene can block programmed cell death (see page 145), suggesting that in cancer, changes that prevent cells from dying can be just as important as those that induce them to divide.

In many tumors, segments of one or more chromosomes are present in many copies. The genes present in this **amplified DNA** are often overexpressed compared with normal cells. If a segment of the genome is often amplified in a particular form of tumor, it is likely that this part of the genome contains a proto-oncogene whose overexpression has contributed to the progression of the tumor. For example, many breast tumors have amplified the *c-myc* proto-oncogene, an early response gene whose product is a potent transcription factor (see page 112). Only some proto-oncogenes can transform cells when they are overexpressed as a result of amplification or translocation. Many others, such as *c-src,* can be converted into active oncogenes only by dominant mutations that alter the structure and regulation of their protein product.

Studies on rare hereditary cancers identify tumor supressor genes

Although transfection experiments showed that many transformed cells carry dominant oncogenes, a different type of experiment revealed that there are also recessive mutations that transform cells by inactivating tumor supressor genes. The existence of such mutations was first suggested by fusing transformed cells to normal cells (Figure 9-5). Many of the hybrid cell lines did not cause cancer in animals, suggesting that the transformed cells harbor recessive mutations that are complemented by wild-type genes from the normal cells. If this hypothesis were true, occasional loss of the chromosomes carrying the complementing genes would lead to reappearance of the transformed phenotype as the population of hybrid cells proliferated. Thus, the recessive transforming mutation could be mapped by determining which chromosome's loss coincides with the reversion to uncontrolled proliferation. Al-

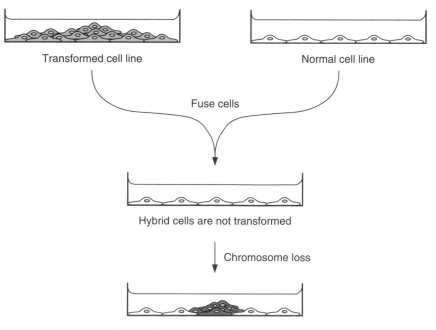

FIGURE 9-5 Hybrids between normal and transformed cells. In many cases fusing a transformed cell to a normal cell produces hybrid cell lines that are not transformed, suggesting that the transformed cell carries recessive mutations that inactivate genes that normally restrain cell growth. The hybrid cell lines can revert to the transformed phenotype by losing the chromosome from the wild type partner that complements the defect in the transformed cell.

Transformed cell line

Normal cell line

Fuse cells

Hybrid cells are not transformed

Chromosome loss

Some cells revert to transformed phenotype

though these studies were provocative, the effects of altering the dosage of normal genes made them complicated to interpret and limited their value as a method for finding genes involved in cancer.

Since mammals are diploid, both copies of the wild-type tumor supressor genes must be inactivated or eliminated to produce a phenotype. This requires either two independent mutations that inactivate the two copies of the gene or a single mutation followed by mitotic recombination or chromosome loss to remove the remaining wild-type copy of the gene (Figure 9-6). Interestingly, mitotic recombination seems to be extremely infrequent in mammals, as if its occurrence were deliberately suppressed to prevent the homozygosis of recessive mutations.

The study of rare hereditary cancers confirmed that recessive mutations in tumor supressor genes can transform cells. Some children inherit the tendency to develop **retinoblastoma,** a childhood cancer of the retina, as a dominant trait located on chromosome 13. Careful analysis of the way in which patients develop retinoblastoma led to a remarkable suggestion: Perhaps the molecular defect that patients inherited was actually a recessive mutation that inactivates one copy of the tumor supressor gene called the *Rb* (retinoblastoma) gene. According to this hypothesis, when the eye begins to develop, all the cells contain one wild-type and one mutant copy of the *Rb*

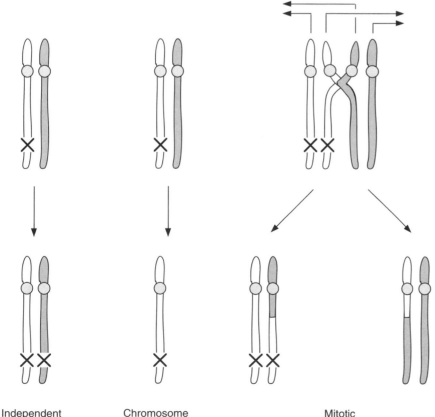

Independent   Chromosome   Mitotic
mutation   loss   recombination

FIGURE 9-6   Different ways of expressing a recessive mutation. Many mutations that alter the growth properties of cells are recessive. After a mutation that inactivates one copy of the gene, several types of events can remove the remaining allele: independent mutations that inactivate the wild-type copy of the gene, deletion of the part of the chromosome that carries the wild-type gene, loss of the chromosome that carries the wild-type copy of the gene (often followed by duplication of the remaining chromosome), or mitotic recombination events that produce cells that have two copies of the original inactivating mutation. Reference: Lasko, D., Cavenee, W., & Nordenskjold, M. *Ann. Rev. Genet.* **25**, 281–314 (1991).

gene. During growth of the retina, a mutation that inactivates the wild-type gene, mitotic recombination, or loss of the chromosome carrying the wild-type copy of the gene could all produce cells that lack a wild-type copy of the *Rb* gene. These cells would lose normal proliferation control and ultimately give rise to retinoblastomas (Figure 9-7).

Although cloning tumor supressor genes is much more difficult than cloning oncogenes, the *Rb⁺* gene was cloned in 1986. As predicted, the normal cells of individuals with the inherited form of retinoblastoma contain both a wild-type and a mutant copy of the retinoblastoma gene, and cells from the retinoblastomas that these individuals develop lack the wild-type copy. Strikingly, the transformation of some cell lines derived from retinoblastoma patients can be reversed by introducing a wild-type copy of the *Rb* gene, providing a direct demonstration of the role of this protein in regulating cell proliferation.

Although individuals with hereditary retinoblastoma develop a very limited spectrum of cancers, the Rb protein is found in all cells. Biochemical experiments suggest that the Rb protein acts as a brake that prevents cells from passing the restriction point by sequestering transcription factors (see pages 114–115). Evidence that the Rb protein has roles outside the retina

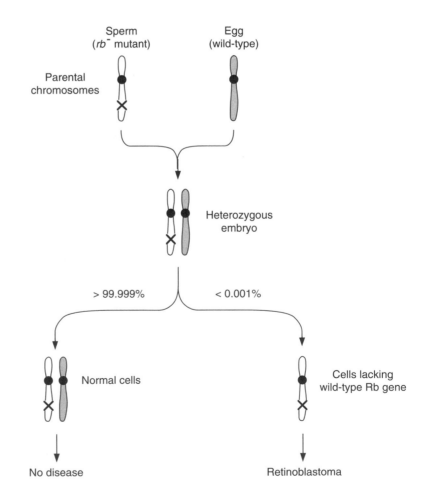

FIGURE 9-7 Retinoblastoma. Retinoblastoma is a rare childhood cancer of the eye. About half of the cases are a result of an inherited mutation that inactivates one copy of the retinoblastoma gene. These heterozygous cells do not give rise to tumors, but rare events that cause the loss of the wild-type copy of the *Rb⁺* gene destroy normal growth controls and initiate the events that produce a tumor.

also comes from crossing pairs of mice that are both heterozygous for a genetically engineered mutation that inactivates the *Rb* gene. A quarter of the progeny from this cross are homozygous for the *Rb⁻* mutation and die during embryogenesis, apparently as a result of defects in blood cell formation.

CANCER AND FEEDBACK CONTROLS

This section discusses how cells accumulate the multiple genetic alterations that give rise to cancer. Cancer cells exhibit many different types of genetic change: point mutations that activate or inactivate genes, amplification of regions of chromosomes, deletions of large parts of chromosomes, missing or extra copies of chromosomes, mitotic recombination events, and translocations.

Identifying proto-oncogenes and tumor supressor genes has made it possible to look for mutations in these genes at different stages of the clinical progression to fully malignant cancer. These studies have led to several important conclusions. First, many different oncogenes and tumor supressor genes are involved in cancer. Second, in a given type of tumor alterations in certain genes are very common, although the detailed pattern of events often varies among different examples of the same tumor. Third, the pathway of events that lead to cancer is different in different types of tumor.

Detailed analysis of colon cancer and glioblastoma (a type of brain tumor) has shown that, as these diseases progress from the first signs of increased cell proliferation to fully malignant tumors, the abnormal cells contain more and more genetic alterations. In colon cancer a small number of genes are mutated in a large fraction of tumors, and certain mutations are usually seen at early stages of the disease while others do not occur until later stages (Figure 9-8). Despite these general principles there are no firm rules for the order or identity of the genetic events: There is no obligatory order in which the changes occur, and certain fully malignant tumors lack any of the changes that are found in the majority of cases. In contrast to colon cancer, the oncogenic changes in glioblastoma appear to occur in a defined order. Comparing the

FIGURE 9-8 Genetic alterations in human cancers. This diagram shows the genetic changes that occur during the progression to malignancy in two human cancers: colon cancer and glioblastoma (a type of brain tumor). The changes are of two types, the inactivation of tumor supressor genes and the activation of proto-oncogenes (boxed). In colon cancer changes occur in a preferred order, but many cases show deviations from this scheme. In glioblastoma the order of changes is believed to be more rigidly defined. The genes shown on the figure are *APC*, adenoma polyposis coli gene identified by studies of families with a genetic predisposition to colon polyps and cancer (function unknown); *ras*, a small G protein that stimulates cell growth (see pages 109–111); *DCC*, deleted in colon cancer, encodes a transmembrane receptor whose function is unknown; *p53* encodes a transcription factor involved in feedback control (see pages 162–165); the γ interferon gene cluster encodes interferons; *erb-B* encodes the epidermal growth factor receptor. The genes altered by the genetic changes on chromosomes 10 and 17 in glioblastoma have not been identified. References: Fearon, E. R., & Vogelstein, B. *Cell* **61,** 759–767 (1990); Lasko, D., Cavenee, W., & Nordenskjold, M. *Ann. Rev. Genet.* **25,** 281–314 (1991).

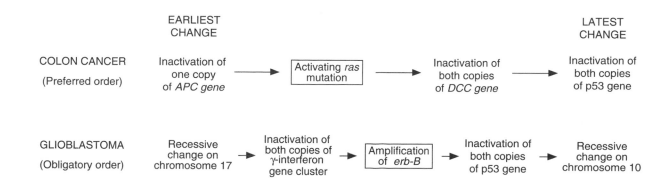

	EARLIEST CHANGE			LATEST CHANGE
COLON CANCER (Preferred order)	Inactivation of one copy of *APC gene*	→ Activating *ras* mutation →	Inactivation of both copies of *DCC gene*	→ Inactivation of both copies of p53 gene
GLIOBLASTOMA (Obligatory order)	Recessive change on chromosome 17	→ Inactivation of both copies of γ-interferon gene cluster → Amplification of *erb-B* →	Inactivation of both copies of p53 gene	→ Recessive change on chromosome 10

mutations seen in colon cancer with those in glioblastoma reveals that only one event, the inactivation of the $p53^+$ gene, is common to both cancers. Mutations that inactivate $p53^+$ are the most frequent genetic change in human cancer and the following section considers the consequences of this mutation.

Mutations that destroy feedback controls may explain the high frequency of cancer

How do humans accumulate the multiple genetic alterations that give rise to cancers? In mammals, the frequency at which one copy of a given gene is inactivated by a spontaneous mutation is about 10^{-6} per cell per generation. Inactivating both copies of a gene requires either two independent mutations or a single mutation followed by loss of the nonmutant copy of the gene by chromosome loss or mitotic recombination. The frequency of these combined events is less than 10^{-8} per cell division. The frequency at which dominant mutations that activate genes arise is in the range of 10^{-8} to 10^{-9} per cell division. Since mutations in several different genes can induce similar alterations in the behavior of cells, we estimate that the frequency of genetic changes that alter the control of cell proliferation, recruitment of blood vessels, or the ability of cells to migrate through the body each occur at a frequency of roughly 10^{-7} per cell division. If the generation of malignant cancer requires four such genetic changes, cancer should arise at a frequency of about $(10^{-7})^4$ or 10^{-28} per cell division. Since there are only 10^{16} cell generations in a human lifetime, cancer should afflict less than one person per billion.

Unfortunately, it is all too obvious that cancer is a common disease. Some of the discrepancy between the predicted and observed incidence of cancer is the result of exposure to carcinogens such as cigarette smoke and sunlight that dramatically elevate mutation frequencies. But we still need other factors to explain why so many people get cancer. One explanation is that there are mutations that have no effect on cell proliferation but that increase the rate at which cells accumulate further mutations. Cancer cells have highly abnormal chromosome complements, offering circumstantial support for the idea that mutations that increase the frequency of genetic damage contribute to the progression of cancer.

Two types of change could increase the rate of subsequent mutations or errors in chromosome segregation. The first are mutations in the DNA replication and repair machinery or components involved in spindle assembly and chromosome segregation. Although such lesions tend to lead to mutations and errors in chromosome segregation, their ability to do so is opposed by feedback controls that detect errors in DNA metabolism or chromosome alignment and give cells time to repair the errors. The second are mutations in the feedback control machinery itself. As we have seen, these mutations greatly increase the frequency of errors in the transmission of genetic information (see page 142).

Analyzing the effects of DNA damage on normal and transformed cell lines reveals that mutations that damage feedback controls play an important role in the development of cancer. Irradiation of mammalian cells has different effects depending on where the cells are in the cell cycle. Cells that are irradiated in G1 cannot pass the restriction point and initiate DNA replication until they have repaired DNA damage. Cells that are irradiated in G2 cannot enter mitosis until the damage has been repaired. Finally, cells that are heavily irradiated during S phase block the firing of replication origins until DNA damage has been repaired.

Although normal cell lines irradiated in G1 fail to enter S phase, many tumor cell lines do not. There is a perfect correlation between cell lines that lack this feedback control and cells that have mutations in the *p53* gene (Figure 9-9). These mutations are of two sorts: recessive mutations that inactivate the gene and dominant mutations that produce abnormal proteins that inhibit the function of the wild-type protein. The results of two experiments confirm that the absence of functional p53 protein causes the lesion in the feedback control: The feedback control can be restored to tumor cells that carry recessive mutations in *p53* by introducing a wild-type *p53+* gene, and the control can be destroyed in normal cells by introducing a dominant *p53D* mutation.

Mutations in *p53* do not affect the feedback control that prevents G2 cells with damaged DNA from entering mitosis. This observation suggests that the p53 protein mediates the ability of signals generated by damaged DNA to block passage through the restriction point, rather than generating the signal that the cell contains damaged DNA. The p53 protein appears to act as a transcription factor that induces the transcription of genes that inhibit progress through the cell cycle. Under normal conditions p53 is an unstable protein and is present at very low levels in the cell. Treating cells with ultraviolet light or X rays dramatically reduces the rate of p53 degradation, leading to a rapid increase in its concentration in the cell and presumably inducing the transcription of genes that block passage through the restriction point.

Studies on gene amplification have convincingly shown that the inactivation of *p53* not only inactivates a feedback control, but also dramatically increases the frequency of genetic damage. Treating rodent cell lines with compounds that inhibit DNA replication can induce the cells to become drug-resistant by amplifying the genes whose products the drugs inhibit. But

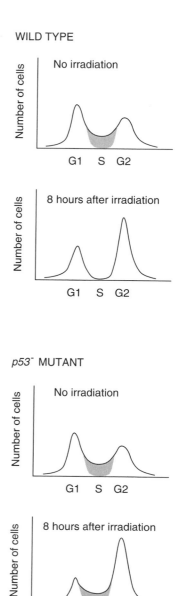

FIGURE 9-9 Feedback controls and p53. FACS (fluorescence-activated cell sorter) profiles of wild-type and *p53−* mutant cells before and 8 hours after irradiation with X rays are shown. In the wild-type cell, irradiation leads to the elimination of S phase cells because damaged DNA prevents passage through the restriction point. In the irradiated *p53−* mutant, however, there is no reduction in the number of cells in S phase, demonstrating that these cells lack this feedback control. The increase in the number of G2 cells in the irradiated *p53−* mutant shows that the feedback control that keeps cells with damaged DNA from entering mitosis is intact in these cells. Reference: Kuerbitz, S. J., Plunkett, B. S., Walsh, W. V., & Kastan, M. B. *Proc. Natl. Acad. Sci. USA* **89**, 7491–7495 (1992).

even though many human tumor cells have amplified genes, attempts to induce gene amplification in normal human cells have been universally unsuccessful, and inhibitors of deoxyribonucleotide synthesis arrest cells in G1. By contrast, human cell lines that lack functional p53 protein enter S phase in the presence of these drugs and show a high frequency of chromosome amplification. Restoring functional p53 to the mutant cells restores the ability of the cell to arrest in G1 and prevents DNA amplification, demonstrating that DNA amplification is in some way a consequence of abnormal DNA replication. The drugs that inhibit deoxyribonucleotide synthesis arrest cells in G1 under conditions in which there should be no breakage. This observation suggests that p53 is part of a system that prevents cells not only from responding to the presence of damaged DNA, but also from entering S phase when they lack the precursors for DNA synthesis.

Some inherited predispositions to cancer inactivate feedback controls

Clear genetic evidence links cancer in humans to lesions in feedback controls. Li–Fraumeni syndrome is an dominant trait associated with an increased frequency of a variety of cancers. As with retinoblastoma, the dominant trait is associated with a recessive mutation, but in this case the mutation lies in the *p53* gene. Relatively frequent events eliminate the wild type copy of the *p53* gene from some of the patient's cells. This change destroys the feedback control that keeps cells with damaged DNA or insufficient DNA precursors from entering S phase and leads to rapid accumulation of genetic damage.

Careful examination of cells from Li–Fraumeni patients suggests that inactivation of a single copy of the *p53* gene produces a subtle but clearly detectable increase in the frequency of cells that contain abnormal chromosome complements. This finding reveals that inactivation of a single copy of certain tumor supressor genes can partially inactivate the feedback controls that maintain the integrity of the genome and thus initiate the events that will give rise to transformation. It remains to be discovered whether the inactivation of other tumor suppressor genes has a similar partially dominant phenotype.

Although the incidence of cancer in individuals with Li–Fraumeni syndrome is higher than average, it is much lower than the incidence of retinoblastoma in individuals who receive only one active copy of the Rb^+ gene. This result makes sense if p53 is a component of a feedback control, whereas Rb is a brake that makes the progress through the restriction point dependent on growth factors. In individuals with only one wild-type *p53* gene, two events are required to destroy normal proliferation control: loss of the wild-type copy of *p53* and a subsequent mutation in some other gene that makes passage through the restriction point independent of growth factors. By contrast, in individuals with only one wild-type *Rb* gene, loss of that wild-type copy is probably sufficient to destroy normal control of cell proliferation. These results suggest a division of labor between two proteins that control passage through the restriction point in response to different signals: p53 is part of a feedback control that responds to damaged DNA, and Rb makes passage through this checkpoint dependent on the presence of growth factors.

CONCLUSION

Understanding the role of feedback controls in cancer has important medical implications. Most therapies against cancer operate on the simple principle that since cells in tumors are actively dividing, agents that kill dividing cells will kill tumor cells. This simple strategy suffers from the obvious limitation that the body contains other dividing cells, such as those that form the lining of the gut and those that give rise to white blood cells. Nevertheless, a largely empirical search for agents that kill tumor cells more effectively than other dividing cells has produced effective strategies for treating several types of tumors. But even when it is successful, chemotherapy is a drastic remedy that usually produces unpleasant side effects both by killing normal dividing cells and by producing abnormalities in nondividing cells. Worse still, there is no effective chemotherapy for many common forms of cancer.

Most chemotherapeutic agents block DNA replication, damage DNA, or interfere with chromosome segregation. Tumor cells probably die because they pass cell cycle checkpoints while downstream events are still inhibited. Thus, the lethal event for cells treated with inhibitors of DNA metabolism would be entry into mitosis, whereas that for cells treated with inhibitors of spindle function would be the onset of anaphase. This argument suggests that treating cells with inhibitors of feedback controls could increase the potency of chemotherapeutic drugs. Killing the same fraction of tumor cells with a lower drug dose would reduce the side effects of chemotherapy that are not due to its effects on cell division. Tests in tissue culture cells and animals have shown that caffeine and related compounds, which keep damaged DNA from preventing entry into mitosis, do increase the killing of tumor cells by agents that damage DNA.

Exploiting differences in the strength of feedback controls in normal and transformed cells might also improve chemotherapy. In mouse cells, feedback controls become weaker as cells become more transformed. Primary cultures show strong feedback controls, established cell lines have somewhat weaker ones, and those of transformed lines are weaker still. By titrating specific inhibitors of feedback controls, it might be possible to abolish the feedback controls of tumor cells, with little effect on those of normal cells. If such conditions could be found they would increase the selective killing of tumor cells, dramatically improving the efficacy of chemotherapy.

We have argued that two very different types of mutation contribute to the genesis of cancer. One type alters the control of cell proliferation, cell death, invasiveness, and other aspects of the relationship of a cell with its neighbors and its environment. The other type makes it easy for cells to accumulate the first type of mutation by impairing their ability to protect themselves from internal errors or external insults that can lead to genetic damage. We expect that during the next few years, analysis of feedback controls in normal cells and studies on tumor suppressor genes will reveal more examples in which lesions in feedback controls play an essential role in the induction of cancer.

SELECTED READINGS

General

Varmus, H., & Weinberg, R. A. *Genes and the Biology of Cancer* (Scientific American Library, New York, 1992). *A comprehensive review of recent advances in the molecular biology of cancer.*

Cairns, J. *Cancer: Science and Society* (W. H. Freeman, New York, 1978). *An excellent examination of the social and medical implications of human cancer.*

Reviews

Lasko, D., Cavenee, W., & Nordenskjold, M. Loss of constitutional heterozygozity in human cancer. *Ann. Rev. Genet.* **25**, 281–314 (1991). *A comprehensive review of the events that inactivate tumor supressor genes in human cancer.*

Cell, **64**, 235–363 (1991). *This issue of the journal Cell contains a series of reviews about different aspects of cancer.*

Frearon, E. R., & Vogelstein, B. A genetic model for colorectal tumorigenesis. *Cell* **61**, 759–767 (1990). *A lucid description of the events involved in the initiation and progression of colon cancer.*

Original Articles

Stehelin, D., Varmus, H. E., Bishop, J. M., & Vogt, P. K. DNA related to the transforming gene(s) of avian sarcoma viruses is present in normal avian DNA. *Nature* **260**, 170–173 (1976). *The first demonstration that viral oncogenes are related to proto-oncogenes found in normal cells.*

Parada, L. F., Tabin, C. J., Shih, C., & Weinberg, R. A. Human EJ bladder carcinoma oncogene is homologue of Harvey sarcoma virus ras gene. *Nature* **297**, 474–478 (1982). *The discovery that mutations in the c-ras proto-oncogene can convert it into an oncogene.*

Knudson, A. G., Jr. Mutation and cancer: Statistical study of retinoblastoma. *Proc. Natl. Acad. Sci. USA* **68**, 820–823 (1971). *The first articulation of the hypothesis that some human tumors are caused by the inactivation of tumor supressor genes.*

Cavenee, W. K., et al. Expression of recessive alleles by chromosomal mechanisms in retinoblastoma. *Nature* **305**, 779–784 (1983). *The demonstration that retinoblastomas arise as a result of recessive genetic changes.*

Kuerbitz, S. J., Plunkett, B. S., Walsh, W. V., & Kastan, M. B. Wild type p53 is a cell cycle checkpoint determinant following irradiation. *Proc. Natl. Acad. Sci. USA* **89**, 7491–7495 (1992). *Analyzing the response of cells to irradiation reveals that p53 is a component of a feedback control that keeps cells with damaged DNA from passing the restriction point.*

THE MEIOTIC
CELL CYCLE

S O FAR, we have discussed "normal" eukaryotic cell cycles, which produce progeny that are genetically identical to their parents. We now turn to the specialized meiotic cell cycle, in which two successive rounds of chromosome segregation follow a round of DNA replication to produce progeny cells with half as many chromosomes as their parents. The function of meiosis appears to vary in different organisms. In multicellular organisms meiosis makes sexual reproduction possible. In unicellular organisms adverse conditions induce the meiotic cell cycle and produce highly resistant spores that can survive much more extreme conditions than cells in the mitotic cell cycle.

The first half of the chapter compares mitotic and meiotic cell cycles to distinguish between processes that are common to all cell cycles and those that are restricted to particular specialized cycles. Chromosome segregation in both meiotic and mitotic cells depends on the interaction of kinetochores and microtubules, but differences in the details of these interactions explain why chromosome segregation in meiosis I differs fundamentally from that in mitosis and in meiosis II.

The second part of the chapter deals with checkpoints in meiotic cell cycles. We have already seen how failure to complete a downstream event, failure to reach a threshold size, or the absence of nutrients or growth factors can arrest cells at cell cycle checkpoints. The genetic analysis of how signals like cell size (Chapter 4), mating factors (Chapter 6), damaged DNA (Chapter 8), and errors in spindle assembly (Chapter 8) regulate passage through checkpoints has been spectacularly successful, but for various reasons these control systems are difficult to analyze biochemically in mitotic cells. In contrast, the biochemical analysis of how oocytes and eggs escape from checkpoints in the meiotic cell cycle has been relatively successful. The meiotic checkpoint can be at G2, metaphase, or G0, depending on the organism. In each case, hormonal stimulation or fertilization can induce large quantities of eggs to move rapidly and synchronously through these checkpoints. We have already seen that the discovery and eventual purification of MPF resulted from studying the control of meiosis in frog eggs (see Chapters 2 and 4).

RECOMBINATION

In almost all eukaryotes, genetic recombination during the meiotic cell cycle contributes to genetic diversity by ensuring that none of the cells produced by the meiotic cell cycle are identical to each other. Recombination is also required for proper chromosome segregation in meiosis I.

Recombination links homologous chromosomes

Diploid cells contain two copies of each chromosome, one inherited from each parent. In outbred populations the sequences of the maternal and paternal copies, or **homologs,** differ by only about 1%. In the meiotic cell cycle DNA replication precedes recombination, so the two homologs that recombine with each other are each composed of a pair of sister chromatids.

How does recombination ensure proper chromosome segregation in meiosis I? The simplest way of thinking about chromosome segregation in the meiotic cell cycle is to begin by ignoring recombination. In meiosis I, the two homologous chromosomes segregate to opposite poles of the spindle, but unlike mitosis, the two sister chromatids that compose each homolog do not separate (Figure 10-1). The meiosis II that follows is essentially identical to mitosis: Sister chromatids separate from each other and segregate to opposite poles of the spindle.

The accurate segregation of chromosomes in mitosis and in meiosis II depends on sister chromatids being physically linked to each other so that they can be oriented toward opposite poles of the spindle. This linkage is probably an inevitable consequence of replication. Likewise, successful segregation of homologs in meiosis I depends on the ability of meiotic recombination to link pairs of homologous chromosomes to each other (Figure 10-2). In a few organisms, like male *Drosophila,* the homologs are linked to each other not by recombination, but by some other, mysterious, linkage.

MITOTIC CELL CYCLE

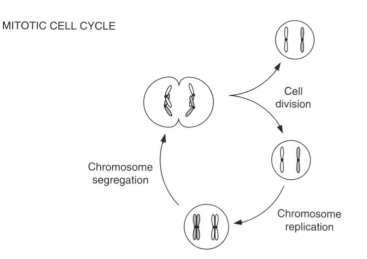

FIGURE 10-1   Meiotic and mitotic cell cycles. The basic features of the mitotic and meiotic cell cycles are shown. In the mitotic cell cycle, DNA replication is followed by a single round of chromosome segregation, in which sister chromatids separate. In the meiotic cell cycle, two rounds of chromosome segregation follow DNA replication. Homologous chromosomes separate from each other in meiosis I, but sister chromatids remain attached and do not separate until meiosis II. Cells do not return to interphase between meiosis I and meiosis II.

MEIOTIC CELL CYCLE

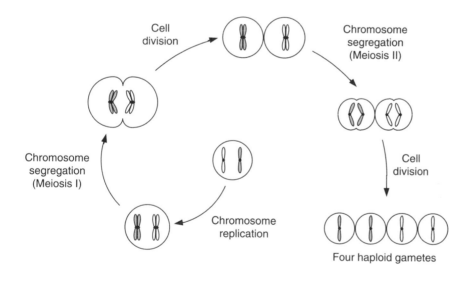

Four haploid gametes

Recombination and synaptonemal complex formation are required for chromosome segregation in meiosis I

What induces meiotic recombination, how does recombination hold homologous chromosomes together, and how is the linkage between these chromosomes dissolved at the end of meiosis I? Although a detailed discussion of meiotic recombination is beyond the scope of this book, we attempt to summarize what is known about these questions.

FIGURE 10-2 Chromosome behavior
in the meiotic cell cycle. In interphase
of the meiotic cell cycle the chromosomes replicate, and homologous chromosomes pair with each other and
recombine. In meiosis I, the homologs
are linked by chiasmata and attach to
opposite poles of the meiotic spindle.
During anaphase I, the arms of the sister chromatids must separate to keep
recombinant chromosomes from being
pulled towards both poles and thus torn
apart, but the sister kinetochores remain firmly linked to each other. In anaphase of meiosis II the sister
kinetochores separate to yield four haploid nuclei. In open meioses the nuclear
envelope breaks down as cells enter
meiosis I and does not reform until the
completion of meiosis II.

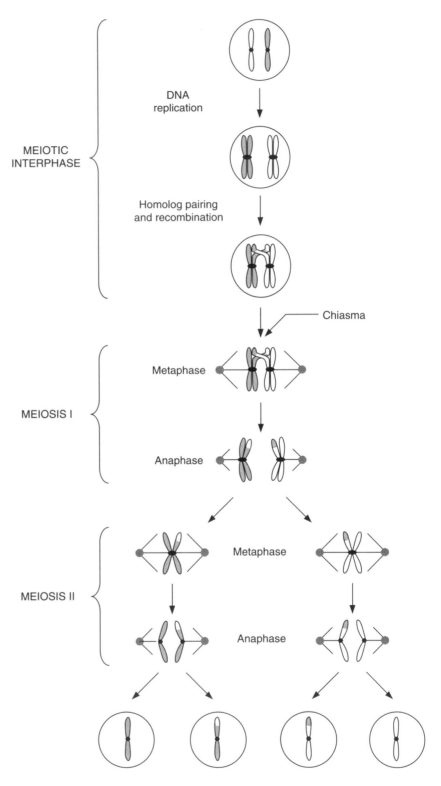

Shortly after DNA replicates in the meiotic cell cycle, the homologs begin to associate, or pair, with each other. Examining cells under the electron microscope suggests that this pairing is accompanied by the formation of a structure called the **synaptonemal complex,** a dense proteinaceous structure that forms along the core of each chromosome (Figure 10-3). Initially each homolog is unpaired and is associated with a part of the synaptonemal complex called the **lateral element.** As the meiotic cell cycle proceeds, the homologs associate with each other so that the two lateral elements lie in register alongside each other. During this process an **axial element** forms between the two lateral elements as if it were zipping the two homologs together. Well before the cells enter meiosis I, the synaptonemal complex disappears, leaving the homologs held together at a small number of points called **chiasmata.**

A variety of approaches have demonstrated that chiasmata correspond to points of **reciprocal recombination** between the homologs. Although its mechanism is complex, reciprocal recombination can be thought of as the product of cutting one chromatid of one homolog and then joining it to a cut chromatid from the other homolog to produce two hybrid chromosomes (Figure 10-4).

In budding yeast there are mutations that separate reciprocal recombination from synaptonemal complex formation. Analyzing these mutants has demonstrated that both reciprocal recombination and axial element formation are required to make sure that homologs segregate in meiosis I. The temporal order of axial element formation and reciprocal recombination has not been clearly established. One popular view is that the initiation of recombination is a prerequisite for axial element formation but that the recombination events are not completed until the synaptonemal complex breaks down.

Homologs are held together by association of sister chromatids

A chromatid that has undergone reciprocal recombination is attached to two different sister chromatids. Between the centromere and the point of recombination, the recombinant is attached to a sister on one homolog, but between the point of recombination and the end of the chromosome, the recombinant is attached to its sister on the other homolog. Since the two homologs separate at anaphase I, the linkage between sister chromatids must dissolve to prevent the recombinant chromosomes from breaking as a result of being dragged toward both poles at the same time (see Figure 10-2). During anaphase of meiosis I, the microscopically visible chiasmata move toward the ends of the chromosomes as if the linkage between the sisters were being progressively dissolved.

We speculate that the synaptonemal complex plays a critical role in maintaining the association of sister chromatids between the time of recombination and anaphase I. Although the visible synaptonemal complex disappears long before metaphase of meiosis I, it may have modified the chromosomes in some way that strengthens the association between the sister

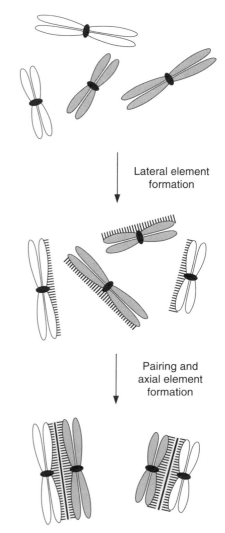

Lateral element formation

Pairing and axial element formation

COMPLETED SYNAPTONEMAL COMPLEXES

FIGURE 10-3 Synaptonemal complex formation. During G2 of the meiotic cell cycle the replicated chromosomes acquire a specialized structure called the synaptomemal complex. First a lateral element forms on each pair of sister chromatids. Then the chromosomes pair with each other and develop an axial element that lies between the two lateral elements. Although the axial element disappears before cells enter meiosis I, its formation is absolutely required to hold the arms of sister chromatids together until anaphase of meiosis I.

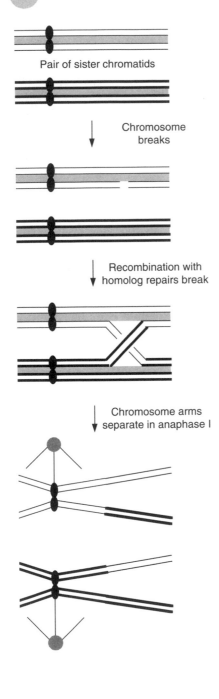

Pair of sister chromatids

Chromosome breaks

Recombination with homolog repairs break

Chromosome arms separate in anaphase I

FIGURE 10-4 Reciprocal recombination. Chromosomes can be linked to each other only by those recombination events that reciprocally exchange regions of two homologous chromosomes. These events are initiated by a double-stranded break on one chromosome. The break is repaired by recombination with the homolog, and about half the time the repair reactions result in the reciprocal exchange diagrammed here. The linkages between sister chromatids are indicated by stippling to emphasize that these linkages must be broken before the homologs can separate in anaphase of meiosis I. Reference: Szostak, J. W., Orr-Weaver, T. L., Rothstein, R. J., & Stahl, F. W. *Cell* **33**, 23–35 (1983).

chromatids. We know nothing about how the linkage between the arms of sister chromatids is broken at the end of meiosis I.

MEIOTIC CHROMOSOME SEGREGATION

A crucial difference in the behavior of sister kinetochores distinguishes meiosis I from mitosis. In mitosis sister kinetochores separate, with the two sister chromatids of each chromosome segregating to opposite poles of the spindle, but in meiosis I, the two sister kinetochores of each homolog do not separate and thus segregate to the same spindle pole. Meiosis II is essentially identical to a mitosis in a haploid cell: The chromosomes have no homologs, and the two sister kinetochores separate.

Kinetochores do not duplicate before meiosis I

What accounts for the unique pattern of chromosome segregation in meiosis I? A variety of evidence supports the idea that regulation of kinetochore behavior underlies the difference between chromosome segregation in meiosis I and that in meiosis II. Detailed cytological studies show that in meiosis I of male *Drosophila,* each homolog carries only one morphologically distinct kinetochore. During anaphase of meiosis I the kinetochore splits into a pair of sister kinetochores, allowing the two chromatids to separate during anaphase of meiosis II. In other organisms the two sister chromatids have visibly distinct kinetochores during meiosis I, but they are constrained to lie side by side, making it difficult for them to attach to opposite poles of the spindle. In mitosis and meiosis II, however, the sister kinetochores face in opposite directions, favoring their interaction with opposite poles of the spindle.

Elegant experiments in which chromosomes are moved between meiosis I and meiosis II spindles also suggest that the difference between meiosis I and meiosis II resides in the chromosomes rather than in the spindles (Figure 10-5). Using a fine glass needle, pairs of homologs from male grasshopper cells were transferred from a meiosis I spindle to a meiosis II spindle. When the meiosis II spindle entered anaphase, the homologs transferred from the meiosis I spindle behaved as they would have done in meiosis I: The two homologs separated, but the sister centromeres of each homolog did not. The

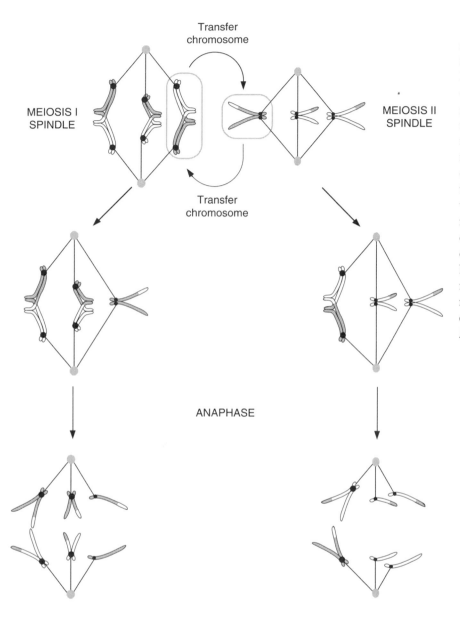

Transfer chromosome

MEIOSIS I
SPINDLE

MEIOSIS II
SPINDLE

Transfer chromosome

ANAPHASE

FIGURE 10-5 Chromosomes determine their own meiotic behavior. Differences between meiosis I and meiosis II chromosomes are responsible for the different patterns of chromosome segregation in meiosis I and meiosis II. Chromosomes were transferred between meiosis I and meiosis II spindles in male grasshopper cells. The meiosis I chromosomes that had been transferred to a meiosis II spindle behaved as they would have done in meiosis I: The homologs separated, but the sister kinetochores did not. Likewise, meiosis II chromosomes on a meiosis I spindle behaved as they would have done in meiosis II: The sister kinetochores separated from each other at anaphase. Reference: Nicklas, R. B. *Phil. Trans. Royal Soc. Series B* **277**, 267–276 (1977).

reciprocal experiment was performed by transferring a pair of chromosomes from a meiosis II spindle to a meiosis I spindle. At anaphase the chromosomes behaved as they would have in meiosis II: The sister centromeres separated, demonstrating that the pattern of chromosome segregation is determined by properties of the chromosomes rather than of the spindles.

The *spo13* mutation specifically blocks meiosis I in budding yeast

Mutations that affect only one of the two divisions emphasize the difference between meiosis I and meiosis II. In budding yeast the *spo13⁻* (sporulation) mutation has no effect on either mitosis or meiosis II but abolishes meiosis I.

The levels of recombination in *spo13⁻* mutants are normal, but the cells form two diploid spores instead of the normal four haploid spores. In contrast to *spo13⁻*, the *cdc31ᵗˢ* mutation, which blocks spindle pole body duplication in the mitotic cell cycle, affects both mitosis and meiosis II but has no effect on meiosis I (Figure 10-6). When *cdc31ᵗˢ* mutants enter the meiotic cell cycle at the restrictive temperature, they form two spores that have completed meiosis I but have not undergone meiosis II.

How does Spo13 ensure that yeast cells perform meiosis I? Recent experiments suggest that Spo13 may induce meiosis I by suppressing events required for entry into mitosis and meiosis II. Spo13 is not normally expressed in the mitotic cell cycle, and experimentally induced expression of Spo13 in

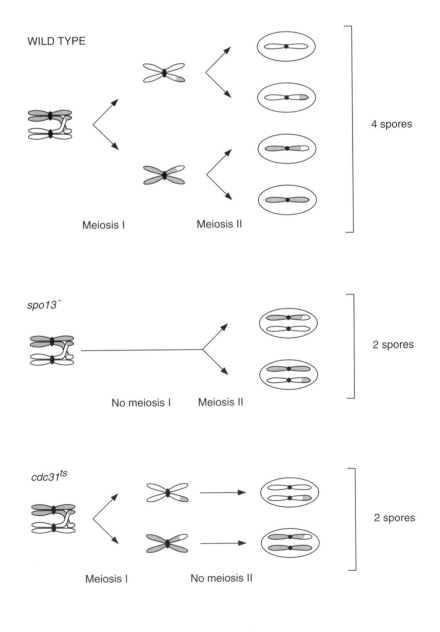

FIGURE 10-6 Mutations that block meiosis I or meiosis II. Different budding yeast mutants fail to perform either meiosis I or meiosis II. The *spo13⁻* mutation blocks meiosis I but allows meiosis II to occur normally. The *cdc31ᵗˢ* mutation prevents spindle pole body duplication at the restrictive temperature and prevents both mitosis and meiosis II but has no effect on meiosis I. Reference: Buckingham, L. E., et al. *Proc. Natl. Acad. Sci. USA* **87**, 9406–9410 (1990).

mitotic cells prevents them from entering mitosis. Perhaps Spo13 is part of a feedback control that keeps cells from entering meiosis II (and its close relative, mitosis) until they have completed meiosis I.

MPF does not regulate chromosome condensation in the meiotic cell cycle

Analysing MPF levels in frog and starfish oocytes as they go through the meiotic cell cycle has raised important questions about the control of downstream events in meiosis. MPF levels fall at the end of meiosis I, when chromosomes segregate and cytokinesis generates the first polar body, but the chromosomes fail to decondense, microtubules do not return to their interphase pattern, and no nuclei form. What prevents these events, which normally are associated with the inactivation of MPF? The most likely answer is that the regulation of protein phosphatases and kinases in the mitotic and meiotic cell cycles differs. The phosphatases that normally reverse MPF-induced phosphorylations could be inhibited, a kinase other than MPF could remain active to phosphorylate some of the substrates normally modified by MPF, or cells could use a combination of these mechanisms.

Members of the MAP family of protein kinases are good candidates for a kinase that keeps cells from entering interphase between meiosis I and II (Figure 10-7). These kinases phosphorylate many of the same sites that MPF does. In addition, MAP kinase activity rises as cells enter meiosis I and remains high between meiosis I and meiosis II, suggesting that this kinase prevents chromosome decondensation and nuclear envelope formation between the two meiotic divisions. Such a scenario may help to explain why the site that MPF phosphorylates in myosin is very different from other MPF phosphorylation sites (see Figure 5-8). Such a nonconsensus phosphorylation site is unlikely to be phosphorylated by MAP kinases, allowing the decline in MPF activity between meiosis I and meiosis II to trigger cytokinesis without a full-blown return to interphase.

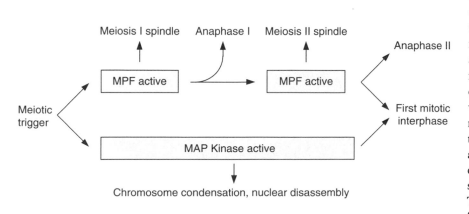

FIGURE 10-7 Model for kinase specialization in meiosis. This speculative model was developed to explain how cells avoid returning to interphase between meiosis I and meiosis II. The periods for which different protein kinases are active are shown. At the end of meiosis I, the cyclin proteolysis machinery is activated, inducing the inactivation of MPF and anaphase I. Between meiosis I and meiosis II, members of the MAP kinase family remain active and phosphorylate nuclear lamins, chromosomal proteins, and other substrates that MPF normally modifies. These phosphorylations prevent the cell from entering interphase until the end of meiosis II.

Meiotic cells do not have to pass Start to begin DNA replication

Budding yeast can switch from the mitotic to the meiotic cell cycle only during G1 (see pages 34 – 37). Why does initiating DNA replication commit cells to the mitotic cell cycle? One possibility is that the specialized pattern of chromosome segregation in meiosis I is the result of differences in the details of DNA replication and kinetochore duplication in the meiotic and mitotic cell cycles. Examining the role of a number of *CDC* genes during the meiotic cell cycle supports this suggestion. In particular, *cdc28^ts* and *cdc7^ts* mutants can replicate their DNA at the restrictive temperature if they are in the meiotic cycle. The function of *CDC7* is required for recombination and spore formation, and *CDC28* function is required for entry into meiosis I only after the completion of DNA replication. These results lead to the surprising conclusion that meiotic DNA replication does not require passage through Start. In this sense the regulation of DNA replication in meiotic cells is similar to that of early embryos.

Although initiating DNA replication commits cells to the mitotic cell cycle, cells that have entered the meiotic cell cycle can return to the mitotic cell cycle at any time before meiosis I. Even cells that have begun recombination can be induced to undergo mitotic chromosome segregation by returning them to rich medium. The observation that commitment to the mitotic cycle is early, whereas commitment to the meiotic cycle is late, suggests that some process that occurs early in the mitotic cycle is delayed until late in the meiotic cycle. Once this process had occurred, cells would be committed to mitotic chromosome segregation, making it pointless to induce other functions involved in recombination and sporulation. Returning cells that are in the

FIGURE 10-8 Model for the meiotic cell cycle. In the mitotic cell cycle, Start induces the duplication of the spindle pole bodies and kinetochores and DNA replication, and mitosis induces sister chromatid segregation. In the meiotic cell cycle there is no Start, and the functional duplication of the kinetochores appears to be suppressed until after the end of meiosis I, allowing the sister kinetochores to segregate to the same pole in meiosis I. An early commitment to kinetochore duplication in the mitotic cell cycle would explain why cells cannot enter the meiotic cycle after the beginning of DNA replication.

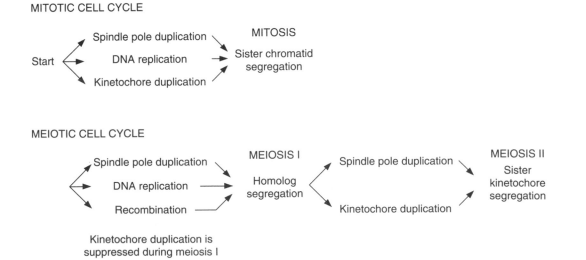

meiotic cell cycle to rich medium would remove the block to this process and induce the mitotic pattern of chromosome segregation. An attractive candidate for this delayed process is the functional duplication of the kinetochores, which commits sister chromatids to segregation and therefore cannot occur before meiosis I (Figure 10-8).

CHECKPOINTS IN MEIOSIS

Eggs and oocytes are excellent material for studying several different cell cycle checkpoints. Organisms that do not protect their offspring produce large quantities of eggs that can be analyzed biochemically. In most of these species, reproductively mature females contain large populations of fully grown oocytes that are arrested at a checkpoint late in G2 of the meiotic cell cycle. These cells are quiescent, with low metabolic rates and little protein synthesis, and can be stored for long periods before maturing into unfertilized eggs. Because the quiescent arrests of G0 cells and G2 cells appear to be similar, understanding how maturation releases oocytes from G2 arrest is likely to help us understand the mechanisms that release somatic cells from quiescent states.

To escape from the G2 checkpoint and become a developing embryo, an oocyte must complete the meiotic cell cycle, activate metabolism and protein synthesis, be fertilized, and begin the mitotic cell cycle. In a few organisms, like clams, the female sheds quiescent, G2-arrested oocytes, and fertilization activates all of these processes. In most other organisms cells are released from the G2 checkpoint only to arrest at a later checkpoint in the meiotic cell cycle before they are fertilized (Figure 10-9). In frogs, secretion of progesterone by the follicle cells that surround the oocyte activates protein synthesis and allows oocytes to progress from their G2 arrest to a second arrest in metaphase of meiosis II. Fertilization releases eggs from this meiotic arrest and starts the mitotic cell cycle. In starfish the secretion of a different, small signal molecule (1-methyladenine) by the follicle cells triggers passage through meiosis I and II, shedding of the oocytes, and metabolic activation. Unless they are fertilized, however, the mature eggs cannot initiate the mitotic cell cycle: They do not replicate their DNA, and even though they contain abundant cyclin B, they do not activate MPF. Despite their evolutionary kinship to starfish, sea urchins use yet another strategy. Their oocytes are induced to mature internally by an unknown stimulus and are retained in the female as unfertilized, metabolically quiescent eggs, arrested in G0. Unlike most eggs, sea urchin eggs can be stored for long periods after maturation before they are shed and fertilized. Fertilization induces the metabolic activation of the egg and starts the mitotic cell cycle. The availability of large quantities of eggs and oocytes from clams, frogs, starfish, and sea urchins makes it possible to biochemically analyze the mechanisms that release cells from G0, G2, and metaphase checkpoints.

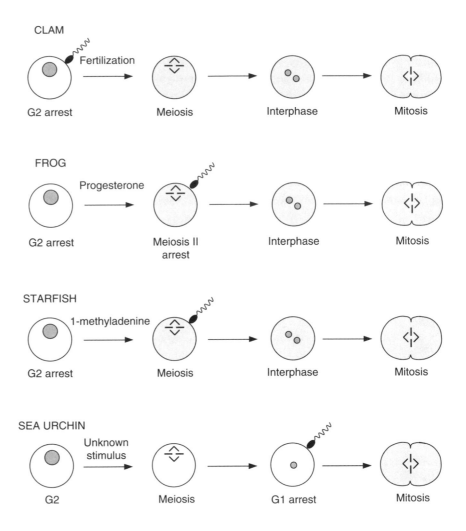

FIGURE 10-9 Oocyte maturation and fertilization. The organization of the last part of the meiotic cell cycle differs in clams, frogs, starfish, and sea urchins. The points at which fertilization and metabolic activation (light stippling) occur are shown. Note that fertilization can release different eggs from G2, metaphase, or G1 arrests—in each case by stimulating an increase in the cytoplasmic calcium concentration.

Meiosis I is induced by activating a preformed pool of preMPF

The release of oocytes from the G2 checkpoint has been studied in frogs, starfish, and clams. In frogs the induction of meiosis I requires new protein synthesis, whereas in starfish and clams it does not. In all three organisms the induction of meiosis II requires protein synthesis. In both frogs and starfish the initial hormonal stimulus for oocyte maturation leads to a reduction of cAMP levels and cAMP-dependent protein kinase activity. Perhaps the oocyte substrates whose phosphorylation arrests cells in G2 are similar to the budding yeast proteins whose phosphorylation keeps cells from entering the meiotic cell cycle.

Several different proteins can induce maturation when injected into frog oocytes. Inhibitors of cAMP-dependent protein kinase induce oocyte maturation by a pathway that requires protein synthesis (Figure 10-10). This dependence suggests that the progesterone-induced reduction in cAMP-dependent protein kinase activity leads to the translation of proteins that induce oocyte maturation. One of these proteins is another protein kinase, the product of the *c-mos* proto-oncogene.

The discovery that c-Mos mRNA exists at high levels in ovaries and testes first suggested that c-Mos plays a role in meiosis. Although present in frog oocytes, c-Mos mRNA is not translated until after oocytes have been induced to begin maturation by progesterone. The importance of c-Mos synthesis in the induction of meiosis was shown by demonstrating that progesterone cannot induce the maturation of oocytes whose c-Mos mRNA has been specifically destroyed using antisense oligonucleotides. It is not clear that c-Mos is the only protein that needs to be synthesized to induce entry into meiosis I.

MPF is activated by removing the inhibitory tyrosine phosphate from a pool of preMPF. In accord with this scheme, preMPF complexes can be isolated from unstimulated oocytes and converted into active MPF in vitro by treating them with purified Cdc25, the tyrosine phosphatase. The initiation of maturation probably reflects a change in the balance between the opposing activities of Cdc25 and Wee1, the tyrosine kinase. In frog oocyte maturation, the synthesis of c-Mos probably plays a key role in increasing the ratio of Cdc25 activity to Wee1 activity (see Figure 10-10); in clams and starfish the balance between these two activities is probably altered by the post-translational activation of MAP kinases.

Induction of meiosis II requires new protein synthesis

The induction of meiosis II is different from that of meiosis I. Proteins like Cdc25 that induce meiosis I cannot induce meiosis II in the absence of protein synthesis. A simple explanation for this result would be that oocytes have to synthesize mitotic cyclins between the end of meiosis I and the onset of meiosis II. Two observations argue against this view, however. First, much of the cyclin B in frog, clam, and starfish oocytes survives the end of meiosis I. Second, ablating the cyclin mRNA in frog oocytes with antisense oligonucleotides fails to prevent meiosis I or meiosis II, showing that oocytes contain all the cyclin that they need to induce both meioses.

The ability of some cyclin to survive meiosis I raises the question of whether cyclin destruction is the event that inactivates MPF at the end of meiosis I. Injecting oocytes with nondegradable cyclin B arrests them in meiosis I, demonstrating that some cyclin must be degraded to allow cells to exit meiosis I. How can we explain the apparently conflicting observations that some cyclin molecules survive meiosis I in order to induce meiosis II, while others must be degraded to inactivate MPF at the end of meiosis I? One possibility is that eggs contain two pools of cyclin. One pool would be associated with Cdc2/28 as preMPF and would be converted into active MPF by Cdc25 as cells enter meiosis I. The activated MPF would induce the degradation of the cyclin in this first pool, leading to the inactivation of MPF and exit from meiosis I. The second cyclin pool would be in some other form, which would not be destroyed at the end of meiosis I. The cyclin in this pool would

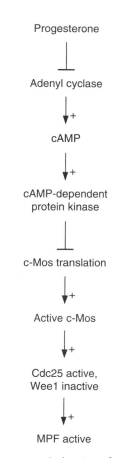

FIGURE 10-10 Induction of meiosis I in frog oocytes. Secretion of progesterone by the follicle cells that surround an oocyte initiates the signaling pathway that leads to meiotic maturation. We do not understand how progesterone leads to the inhibition of adenyl cyclase, how reducing the activity of cyclic AMP-dependent protein kinase induces the translation of c-Mos mRNA, or how the protein kinase activity of c-Mos activates Cdc25. Reference: Yew, N., Mellini, M. L., & Vande Woude, G. F. *Nature* **355**, 649–652 (1992).

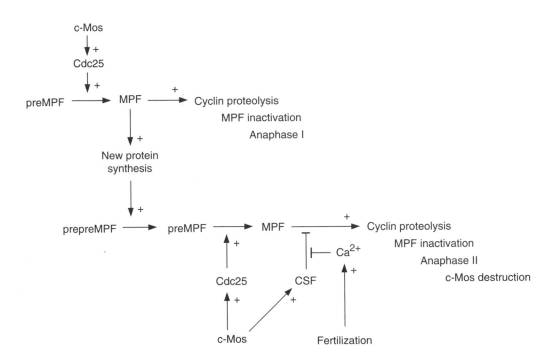

FIGURE 10-11   Regulation of meiosis I and II in frog oocytes. There appear to be distinct pools of cyclin B and Cdc2/28 in frog oocytes that are activated in meiosis I and meiosis II. In meiosis I, the first pool (preMPF) is converted into active MPF by removal of phosphotyrosine that is catalyzed by Cdc25 that has been activated as a result of the translation of c-Mos mRNA. A second pool, which we call prepreMPF, is activated by poorly characterized reactions that require the synthesis of other new proteins in addition to c-Mos. In meiosis II, the presence of c-Mos leads to cytostatic factor (CSF) activity that blocks the ability of active MPF to induce cyclin degradation and to thus trigger the inactivation of MPF.

require the synthesis of other proteins, in addition to c-Mos, to allow it to form active MPF (Figure 10-11).

Cytostatic factor (CSF) arrests unfertilized frog eggs in metaphase of meiosis II

Unfertilized frog eggs are arrested at metaphase of meiosis II. Fertilization or other treatments that increase the intracellular calcium level release cells from metaphase arrest by inducing cyclin degradation and MPF inactivation. A **cytostatic factor (CSF)** responsible for the meiotic arrest was discovered at the same time that MPF was. CSF was identified by transferring cytoplasm from an unfertilized eggs to fertilized eggs. The injected cells arrested in metaphase, demonstrating that the donor unfertilized eggs contained some substance (CSF) that can arrest the cell cycle (Figure 10-12).

CSF arrests the cell cycle by inhibiting cyclin degradation. Unlike MPF, CSF has never been purified, so we are not certain of its composition. The results of two experiments suggest that c-Mos plays a key role in the generation and maintenance of CSF activity. First, c-Mos synthesis after the end of meiosis I is required both for the activation of MPF that induces meiosis II and for the cell cycle arrest in metaphase of meiosis II. Second, antibodies to c-Mos can deplete CSF activity from unfertilized egg cytoplasm, demonstrating that this protein is either a component of CSF or plays an essential role in maintaining the activity of CSF.

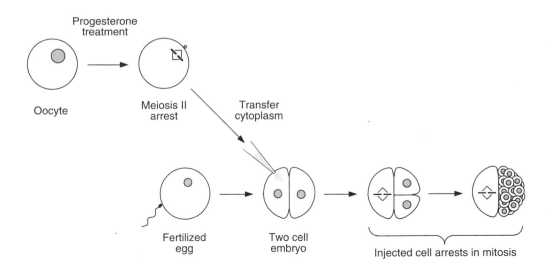

Fertilization-induced increases in calcium concentration release eggs from many different checkpoints

We have seen that unfertilized eggs arrest in G2 of the meiotic cell cycle in clams, in metaphase of meiosis II in frogs, and in G1 of the first mitotic cell cycle in sea urchins. Although their eggs arrest at different checkpoints, all of these organisms use a fertilization-induced increase in the intracellular calcium concentration to restart the cell cycle. Thus, the same signal can release cells from a variety of checkpoints. One interpretation of this finding is that increased calcium is a universal signal used to release cells from any checkpoint. The observation that experimentally elevating intracellular calcium releases clam oocytes from their G2 arrest but cannot induce G2-arrested frog or starfish oocytes to enter meiosis argues against this hypothesis. A more likely possibility is that the different eggs have each harnessed the reactions induced by calcium to overcome a different cell cycle checkpoint. This question will not be resolved until we understand the molecular mechanisms by which calcium releases cells from arrest.

FIGURE 10-12 Discovery of cytostatic factor (CSF). Cytoplasmic transfer experiments revealed the existence of CSF. Cytoplasm removed from unfertilized eggs and injected into one cell of a two-cell embryo induced the injected cell to enter a prolonged metaphase arrest at the next mitosis. This experiment showed that the unfertilized eggs contained some factor capable of inducing cell cycle arrest. Reference: Masui, Y., & Markert, C. L. *J. Exp. Zool.* **177,** 129–45 (1971).

CONCLUSION

Studies of the meiotic cell cycle are at a frustrating stage. They raise important questions about the cell cycle but cannot yet answer them. The analysis of meiosis I suggests that control of kinetochore behavior can regulate the pattern of chromosome segregation, but we do not know how this behavior is regulated. The ability of cells at the end of meiosis I to avoid destroying all their mitotic cyclins suggests that there can be two pools of cyclin in the same cell, but we do not know the biochemical differences between these pools. Understanding this phenomenon is likely to offer clues about how mitotic cells that constitutively express G1 or mitotic cyclins still show an orderly alternation of Start and mitosis. Likewise, the failure of the fall in MPF activity between meiosis I and II to induce the events that lead to interphase explains how cells block the initiation of replication between the two meioses. Unfortunately, we do not understand how cells maintain chromosome condensation and suppress nuclear re-formation in the absence of MPF. Finally, although eggs and oocytes are excellent subjects for studying cell cycle checkpoints, we do not yet know the details of the signaling pathway that leads from any stimulus to the resumption of the cell cycle.

SELECTED READINGS

General

Moens, P. B., ed. *Meiosis* (Academic Press, San Diego, 1987). *A full review of various aspects of meiosis.*

Kucherlapati, R., & Smith, G. R. *Genetic recombination.* (American Society for Microbiology, Washington, DC, 1988). *A collection of reviews on various aspects of recombination.*

Reviews

Roeder, G. S. Chromosome synapsis and genetic recombination: their roles in meiotic chromosome segregation. *Trends in Genet.* **6,** 385–389 (1990). *A review of recent evidence that recombination may be a prerequisite for synapsis.*

Masui, Y., & Clarke, H. J. Oocyte maturation. *Int. Rev. Cytol.* **57,** 185–282 (1979). *A comprehensive, although somewhat dated, review of the events involved in inducing meiosis in oocytes.*

Original Articles

Szostak, J. W., Orr-Weaver, T. L., Rothstein, R. J., & Stahl, F. W. The double-strand-break repair model for recombination. *Cell* **33,** 23–35 (1983). *The proposal that meiotic recombination is initiated by double-stranded breaks in DNA.*

Malone, R. E., & Esposito, R. E. Recombinationless meiosis in *Saccharomyces cerevisiae. Mol. Cell. Biol.* **1,** 891–

901 (1981). *Analysis of a mutant that does not perform meiosis I reveals that recombination is required only for correct chromosome segregation in meiosis I.*

Nicklas, R. B. Chromosome distribution: Experiments on cell hybrids and in vitro. *Phil. Trans. Royal Soc. Series B* **277,** 267–276 (1977). *Micromanipulation experiments show that the difference between meiosis I and meiosis II lies in the chromosomes rather than the spindles.*

Rockmill, B., & Roeder, G. S. Meiosis in asynaptic yeast. *Genetics* **126,** 563–574 (1990). *Analysis of a mutant that allows recombination but prevents synaptonemal complex formation shows that recombination alone cannot ensure proper chromosome segregation in meiosis I.*

Sagata, N., Watanabe, N., Vande Woude, G. F., & Ikawa, Y. The c-mos proto-oncogene product is a cytostatic factor responsible for meiotic arrest in vertebrate eggs. *Nature* **342,** 512–518 (1989). *The c-Mos protein is shown to be initimately associated with the cytostatic factor activity responsible for arrested unfertilized frog eggs in metaphase of meiosis II.*

Shuster, E. D., & Byers, B. Pachytene arrest and other meiotic effects of the Start mutations in *Saccharomyces cerevisiae. Genetics* **123,** 29–43 (1989). *Evidence that Cdc28 is not required for DNA replication in the meiotic cell cycle.*

THE BACTERIAL
CELL CYCLE

*L*IKE EUKARYOTES, bacteria display an enormous amount of diversity in their appearance and the organization of their cell cycles. This chapter concentrates on the bacterium *Escherichia coli,* which is an excellent subject for studying the cell cycle. *E. coli* divides rapidly on simple media, can be obtained in large quantities, and has been subjected to intensive genetic and biochemical analysis. Because of these advantages, many processes, including the mechanism of DNA replication, are much better understood in *E. coli* than they are in any eukaryote. Surprisingly, although many bacterial mutants affect the processes of DNA replication and cell division, no mutants like the budding yeast *cdc28^ts* mutant, which arrests all processes in the cell cycle, have been found. As a consequence, much of our knowledge about the overall organization of the bacterial cell cycle comes from physiological experiments, which tend to emphasize the differences rather than the similarities between bacterial and eukaryotic cell cycles. Throughout this chapter we attempt to assess the similarities between the organization of the bacterial and eukaryotic cell cycles and the strategies that they use to solve the universal problems of the cycle.

ORGANIZATION OF THE BACTERIAL CELL CYCLE

Bacteria are very different from eukaryotic cells. Bacteria lack intracellular organelles, most notably the nucleus, and the logical organization of their cell cycles seems quite different from that of eukaryotes. The significance of the differences between bacterial and eukaryotic cell cycles is dificult to assess. At one extreme, the fundamental rules and molecules that govern the bacterial cell cycle might be completely different from those that govern the eukaryotic cycle. At the other, the apparent differences between the two cycles, like those between early embryonic and somatic cell cycles, might conceal similar organizing principles. Because eukaryotes and bacteria are believed to have descended from a common progenitor, we favor the idea that fundamental similarities exist between their cell cycles.

The organization of bacterial cells differs from that of eukaryotes

At first glance, *E. coli* looks like a miniature version of fission yeast. The rod-shaped cells grow by elongating and divide by forming a septum across the middle of the rod. Closer inspection, however, reveals that the basic architecture of this bacterium is very different from that of any eukaryote (Figure 11-1). Unlike eukaryotes, bacteria have no internal membrane-bounded organelles. An inner and outer membrane separate the contents of the cell from the environment. Between these membranes lies a rigid cell wall composed of **peptidoglycans,** which are extensively cross-linked polymers made up of short peptides and sugars.

The bacterial genome is smaller and more simply organized than that of eukaryotes. The multiple linear chromosomes of eukaryotes reside within the nucleus, whereas prokaryotes have a single small circular chromosome organized into a structure called the **nucleoid** that is in direct contact with the cytoplasm. The bacterial chromosome has a single highly specialized replica-

FIGURE 11-1 Structure of a bacterial cell. This view of *Escherichia coli,* the most intensively studied bacterium, shows that bacteria have no discrete organelles and no membrane to separate the nucleoid, which is composed of the chromosomal DNA, from the cytoplasm. The cell is bounded by an inner and outer membrane, which are separated by a cell wall composed of peptidoglycans. The polar and periseptal annuli are specialized sites where the inner and outer membranes come into close contact with each other.

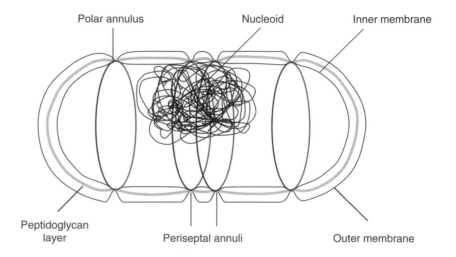

tion origin, while eukaryotic chromosomes each have multiple origins. For at least part of the cell cycle, the bacterial replication origin appears to be attached to the internal surface of the inner membrane. The *E. coli* chromosome is 4.5 million base pairs in length, one-third the size of the genome of budding yeast and one-thousandth the size of the human genome. Bacteria lack any of the cytoskeletal systems that move organelles within eukaryotic cells. Attempts to visualize an organized structure that segregates the chromosomes have not yet succeeded.

A round of bacterial DNA replication can take longer than the cell cycle

Like eukaryotes, prokaryotes must coordinate chromosome replication, chromosome segregation, and cell division to ensure the faithful transmission of genetic information. At first sight, the bacterial solutions to these problems seem radically different from those of eukaryotes. These apparent differences mostly stem from one striking fact: The cell cycle of rapidly growing bacteria is much shorter than the time required to complete a round of DNA replication and chromosome segregation. On rich medium, *E. coli* cells divide every 20 minutes, but it takes a minimum of 40 minutes to complete a round of DNA replication and another 20 minutes to segregate the replicated chromosomes and divide. There is only one way to solve the problems created by this seemingly bizarre arrangement: During rapid growth, rounds of DNA replication must begin in one cell cycle and finish in another.

Figure 11-2 shows the pattern of DNA replication in an *E. coli* population growing with a doubling time of 35 minutes. A newborn cell contains a chromosome that already has a pair of replication forks moving around the chromosome. Ten minutes after birth the two replication origins that lie behind these forks fire synchronously to initiate the next round of replication. Shortly afterward, the older forks meet to terminate the round of replication that was initiated in the preceding cell cycle and the two chromosomes segregate from each other, even though each chromosome is engaged in a round of replication that it will not complete until after cell division. Therefore, in rapidly growing bacteria DNA replication occurs throughout the cell cycle. Only in cells that have doubling times of 60 minutes or more is each round of DNA replication initiated and terminated in the same cell cycle, resulting in a cell cycle that looks more like that of eukaryotes.

Bacterial cells have a threshold mass for initiating DNA replication and a threshold length for chromosome segregation

The continuous DNA replication in rapidly growing bacteria obscures the similarities between prokaryotic and eukaryotic cell cycles. Careful analysis, however, reveals that, like eukaryotic cells, bacteria must reach threshold sizes to trigger DNA replication and chromosome segregation. In eukaryotes size thresholds were revealed by the analysis of mutants; in bacteria their

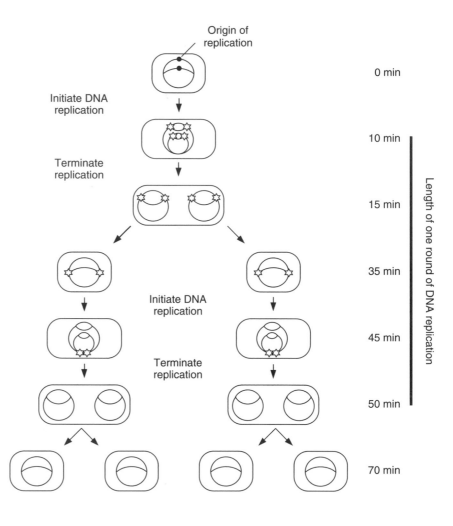

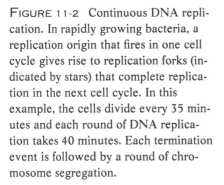

FIGURE 11-2 Continuous DNA replication. In rapidly growing bacteria, a
replication origin that fires in one cell
cycle gives rise to replication forks (indicated by stars) that complete replication in the next cell cycle. In this
example, the cells divide every 35 minutes and each round of DNA replication takes 40 minutes. Each termination
event is followed by a round of chromosome segregation.

existence was deduced from physiological analysis of cells growing at different rates.

Under a wide range of conditions, the mass at which a cell initiates DNA
replication divided by the number of replication origins the cell contains is a
constant, called the **initiation mass.** Therefore, rapidly growing cells, which
contain multiple replication origins, initiate DNA synthesis at a larger mass
than do slowly growing cells, which contain only a single origin. This behavior is analogous to that of diploid yeast cells, which must reach a larger mass
than haploids have to before they can pass Start.

Eukaryotic cells must also attain a critical size before they can enter
mitosis and segregate their chromosomes (see pages 44–46). A simple experiment demonstrated that bacterial cells have to reach a critical length before
they can segregate their chromosomes and divide. In *E. coli,* inhibitors of
protein synthesis block cell growth and initiation of DNA replication but do
not block the elongation or termination stages of replication. Therefore,
treating cells with a protein synthesis inhibitor allows rounds of replication
that have already begun to finish. When an exponentially growing population
of cells was treated with a protein synthesis inhibitor, all the cells that had

already segregated their chromosomes into two discrete nucleoids divided. The other cells completed DNA replication but did not segregate the daughter chromosomes, even though these chromosomes had terminated replication and were no longer topologically linked. When the cells were allowed to resume protein synthesis, the replicated chromosomes did not segregate until the cells had exceeded a threshold length. When cells reached this length, the chromosomes segregated rapidly to new positions one-quarter and three-quarters along the length of the cell. The threshold length for chromosome segregation is precisely twice the smallest length at which slowly growing cells are born.

Can cell length really control cell behavior? One clue came from observing cells grown at different rates: At the time of chromosome segregation the cell length was always the same, whereas the cell mass varied with the growth rate. Studying the division of cells of different shapes provided a striking demonstration that cell length regulates chromosome segregation and cell division. Normal *E. coli* cells are rod-shaped, but mutants in cell wall synthesis can produce elliptical or even spherical cells. The initiation mass and critical length for chromosome separation is the same for all three cell shapes (Figure 11-3), but the mass at chromosome segregation increases dramatically as cell shape changes from rod to ellipsoid to sphere. This result demonstrates that cell length, rather than cell mass, determines when chromosome segregation occurs.

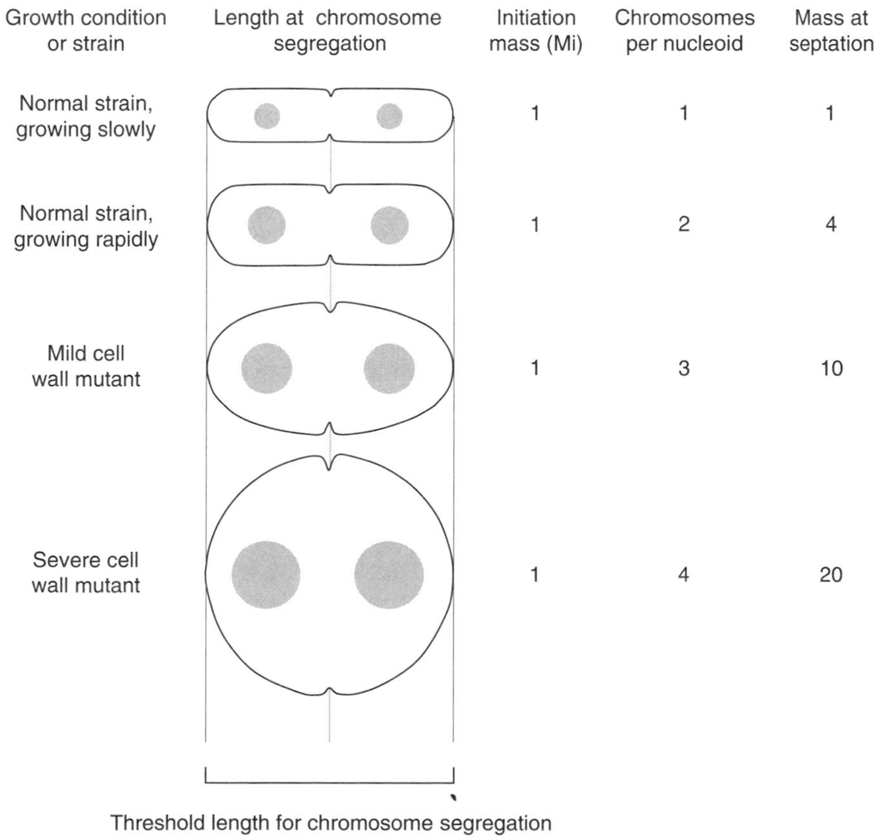

Growth condition or strain	Length at chromosome segregation	Initiation mass (Mi)	Chromosomes per nucleoid	Mass at septation
Normal strain, growing slowly		1	1	1
Normal strain, growing rapidly		1	2	4
Mild cell wall mutant		1	3	10
Severe cell wall mutant		1	4	20

Threshold length for chromosome segregation

FIGURE 11-3 Cell length controls chromosome segregation. The critical sizes for initiation of DNA replication and chromosome segregation are compared in four populations of *E. coli:* slowly growing normal cells, rapidly growing normal cells, elliptical mutant cells, and spherical mutant cells. The initiation mass (Mi) is the cell's mass at the initiation of DNA replication divided by the number of replication origins the cell contains. Note that initiation mass is independent of cell shape. In contrast, although the mass at the time of chromosome segregation varies widely, the length of the cells at segregation is independent of shape, suggesting that it is cell length rather than cell mass that regulates chromosome segregation. Reference: Donachie, W. D., & Begg, K. J. *J. Bacteriol.* **171,** 4633–4639 (1989).

The organization of the bacterial cell cycle is similar to that of eukaryotes

Does the prokaryotic bacterial cell cycle have discrete transitions that correspond to Start, entry into mitosis and exit from mitosis in eukaryotic cells? Although this question has no definitive answer, a comparison of the two types of cell cycle suggests that they may have certain underlying similarities.

Eukaryotic cells must reach a critical mass to pass Start and initiate DNA replication; and bacteria must exceed the initiation mass to begin replication. Thus, we can regard attaining the initiation mass in bacteria as inducing an event similar to Start and the interval between the beginning of the bacterial cell cycle and initiation of replication as corresponding to G1 in eukaryotes. Toward the end of their cell cycle, eukaryotes reach a critical size, activate MPF, and enter mitosis; prokaryotes bacteria reach a critical length, triggering chromosome segregation and ultimately cell division. If we regard chromosome segregation in bacteria as analogous to mitosis in eukaryotes, the interval between termination of a round of replication and chromosome separation in bacteria is equivalent to the eukaryotic G2. When cells grow rapidly, the bacterial equivalents of G1 and G2 are not seen as gaps between periods of DNA replication because the S phases of successive cell cycles overlap.

In eukaryotes, each origin of DNA replication fires once per cell cycle, and there is a rigid alternation between Start and mitosis. Bacteria growing at a constant rate also initiate replication exactly once per cell cycle. Slowly growing cells are born with a single replication origin that fires once during each cell cycle, while rapidly growing cells are born with as many as four origins, which all fire synchronously once per cell cycle. There is, however, no obligatory alternation between the initiation of replication and bacterial cell division. When rapidly growing cells are switched to slow growth conditions, cell cycles in which origins do not fire at all must occur so that the number of origins per cell can be reduced from four to one. Conversely, when cells are switched from slow to rapid growth conditions, there must be transitional cells in which the origins fire more than once per cell cycle so that the number of origins per cell can increase to more than one (Figure 11-4).

The absence of tight coupling between the initiation of replication and chromosome segregation in bacteria may reflect the existence of two separate oscillators that control the bacterial cell cycle (Figure 11-5). One of these oscillators would periodically initiate DNA replication and would be restrained from doing so until cells had reached the initiation mass. The other oscillator would induce chromosome segregation and would be restrained from doing so until cells had achieved the critical length for chromosome segregation. When cells are growing at a constant rate, they alternately reach the initiation mass and the threshold length for chromosome segregation, so rounds of initiation and segregation do alternate with each other. There appear to be special mechanisms that help to break this regular alternation when cells switch from one growth rate to another.

The somatic cell cycle engine in eukaryotes can also be thought of as two oscillators. In one oscillator complexes between G1 cyclins and Cdc2/28 induce Start, and in the other, complexes between mitotic cyclins and Cdc2/

STEADY-STATE CYCLES

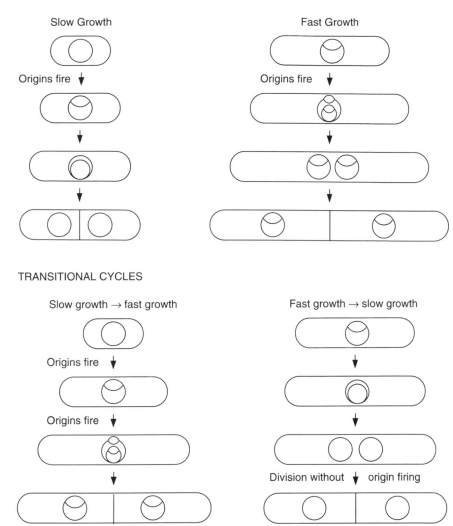

FIGURE 11-4 Coordination of chromosome replication and cell division. *E. coli* growing under steady-state conditions, whether slowly or quickly, fire their replication origins once per cell cycle. When shifted to richer medium, cells grow faster, and a second round of replication is initiated before the cells divide (bottom left). Conversely, when shifted from a rich to a poor medium, cells are born with a replication fork already in progress and do not initiate a new one until after the next cell division (bottom right).

28 induce mitosis. But, unlike those of bacteria, these oscillators are tightly coupled so that Start and mitosis always alternate with each other. We speculate that eukaryotes evolved the coupling between the Start and mitosis oscillators when chromosome condensation arose. If eukaryotes attempt to condense replicating chromosomes, they suffer severe DNA damage, mandating a strict alternation between replication and segregation. In bacteria, however, the absence of condensation means that replicating chromosomes can be segregated without being damaged. The idea that two independent oscillators control the bacterial cell cycle may explain why it has been possible to find bacterial cell cycle mutants that can block either DNA synthesis or cell division, but not both. By contrast, the coupling between chromosome replication and segregation in eukaryotes means that most cell cycle mutants arrest both replication and segregation within a single cell cycle.

Do the molecules that regulate the bacterial cell cycle bear any similarity to the components of the eukaryotic cell cycle engine? None of the identified

FIGURE 11-5 Prokaryotic and eukaryotic cell cycle engines. Prokaryotes appear to have two independent oscillators that control DNA replication and chromosome segregation. The need to reach a critical cell length restrains the oscillator that controls chromosme segregation, and the need to reach the initiation mass (Mi) restrains the oscillator that controls DNA replication. Eukaryotic cells can also be thought of as having two connected oscillators, one that induces mitosis and one that induces Start. Each oscillator depends on the progress of the other, thus, passage through mitosis allows the Start oscillator to initiate another cycle. Likewise, passage through Start allows the mitotic oscillator to initiate a new cycle. Both oscillators are also restrained by the need to reach a threshold size.

PROKARYOTES: TWO INDEPENDENT CELL CYCLE ENGINES

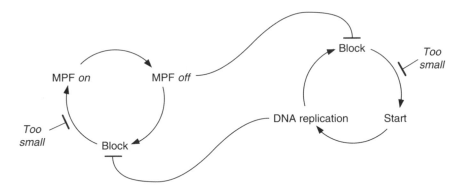

EUKARYOTES: TWO CONNECTED CELL CYCLE ENGINES

genes that control DNA replication or cell division in bacteria show any relationship to cyclins or Cdc2/28. Moreover, unlike eukaryotes, bacteria have very few protein kinases that regulate the activity of their substrates by phosphorylating them on serine and threonine residues. Nevertheless, the identification of an increasing number of conserved proteins that perform the same function in prokaryotes and eukaryotes suggests that we should not dismiss the possibility that bacteria have homologs of cyclins and Cdc2/28.

DNA REPLICATION

DNA replication in *E. coli* is probably the best understood cell cycle event in any organism. Our knowledge comes from a combination of genetic analysis and a concerted effort to identify the components of the DNA replication machinery, purify them, and get them to replicate DNA in a test tube. This project has taken more than 20 years and reveals the discipline and determination that will be needed to unravel the molecular details of how cell cycle engines work.

DnaA is critical for the initiation of DNA replication

Molecular studies have identified the *E. coli* DNA replication origin and the proteins that interact with it. The replication origin is 245 base pairs long and contains a group of four binding sites for a protein called **DnaA** that are located near three repeats of a short adenine/thymine-rich DNA sequence (Figure 11-6). The firing of the origin is dependent on the binding of many molecules of DnaA that carry ATP, although the exact role of nucleotide binding and hydrolysis in activating the origin is unclear. The activity of DnaA is altered when it binds to phospholipids, suggesting that binding of the replication origin to the cell membrane may play a role in regulating DNA replication.

The formation of a large protein–DNA complex activates the chromosomal replication origin. About 30 molecules of DnaA and 5 molecules of a bacterial histonelike protein bind to the origin region, forming a nucleosome-like structure in which a loop of DNA is wrapped around the complex of proteins (Figure 11-7). The bound proteins trigger the progressive unwinding of the A/T-rich DNA repeats until about 45 nucleotides have become single-stranded. A DNA-binding protein coats the unwound DNA strands to keep them from reannealing, while the bound DnaA recruits a DNA helicase that can move along the chromosome, unwinding the DNA in front of the replication fork. The helicase in turn recruits a DNA primase to the growing replication complex. The primase synthesizes the short RNA molecules that prime DNA synthesis, which is catalyzed by a multisubunit DNA polymerase. Once started, the polymerase molecules move away from the replication fork in both directions, synthesizing DNA at the astonishing rate of 1,000 nucleotides per second. DNA topoisomerases function as swivels to prevent the template DNA from becoming tangled as it unwinds. Comparing the steps of initiation in prokaryotes with those in the eukaryotic virus SV40 (see Chapter 7) reveals that the basic mechanisms for initiating DNA replication are remarkably similar and that the sequences of prokaryotic and eukaryotic DNA polymerases and accessory factors show clear homology to each other.

What determines the initiation mass? Attempts to answer this question have yielded confusing results, partly because two events appear to be required for a complete round of replication. The first is the firing of the replica-

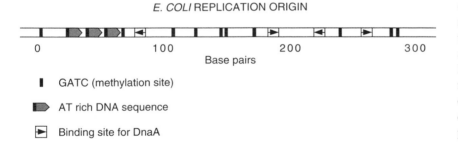

E. COLI REPLICATION ORIGIN

0   100   200   300
Base pairs

▮ GATC (methylation site)

▰▷ AT rich DNA sequence

▷ Binding site for DnaA

FIGURE 11-6 The bacterial replication origin. The *E. coli* replication origin contains three different types of element. There are four binding sites for the DnaA protein and three repeats of an A/T-rich DNA sequence. Binding of DnaA triggers the unwinding of the A/T-rich sequences. In addition, the origin contains a number of repeats of the sequence GATC, whose methylation plays a role in blocking rereplication.

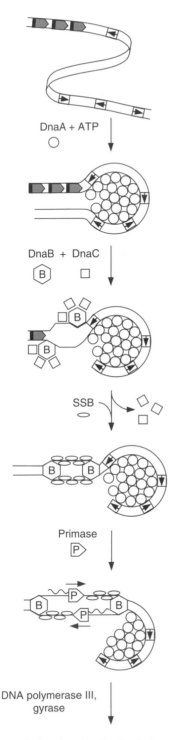

DnaA + ATP

DnaB + DnaC

SSB

Primase

DNA polymerase III,
gyrase

Bidirectional replication fork

FIGURE 11-7 Initiation of DNA replication. The binding of DnaA molecules (carrying bound ATP) to the replication origin leads to the unwinding of the A/T-rich repeats in the replication origin. A complex of DnaB and DnaC binds to the unwound DNA, and the helicase activity of DnaB extends the unwound region, which is prevented from reannealing by binding of a single-stranded binding protein (SSB). A primase binds to the unwound DNA and synthesizes the short RNA molecules that are the primers for DNA synthesis by DNA polymerase III. DNA gyrase is a prokaryotic type II DNA topoisomerase.

tion origin, which seems to be regulated by the concentration of DnaA in the cell. The second is a very poorly characterized control that seems to block the elongation of the replication forks if cells are too small.

The mechanism by which DnaA controls replication is unclear. The concentration of DnaA is considerably higher than the level needed for the initiation complex to form in vitro and does not appear to fluctuate during the cell cycle. Although suboptimal expression of DnaA increases the initiation mass, overexpressing the protein from a heterologous promoter hardly decreases it at all. The inability to dramatically reduce the initiation mass may reflect the control that can arrest the elongation phase of replication. Identifying the additional factors that link the early stages of DNA replication to cell mass remains an important goal for future research.

The block to rereplication involves DNA methylation and membrane attachment of the replication origin

What prevents a bacterial replication origin from firing twice in rapid succession? In bacteria, there is no boundary between chromosomes and cytoplasm, so prokaryotes cannot use the licensing factor mechanism found in eukaryotes (see pages 127–129). One clue as to how bacteria regulate the frequency of origin firing came from studies that revealed that, in rapidly growing cells, all the origins fire at the same time. This synchrony is lost in *dam⁻* (deficient in adenine methylase) strains, which are unable to methylate the adenine residues in the DNA sequence GATC, which occurs several times in the origin region (Figure 11-8). Immediately after replication, DNA sequences are methylated on the parental DNA strand but not on the newly synthesized strand (Figure 11-9). When in this hemimethylated state, the replication origin binds to specific regions of the cell membrane that are believed to inhibit its rereplication. Consistent with this idea, hemimethylated plasmids that carry the chromosomal replication origin cannot replicate when they are introduced into *dam⁻* strains (Figure 11-10). Interactions with the membrane may also be important for controlling the activity of DnaA since phospholipids induce the protein to exchange bound ADP for ATP and to form aggregates.

Although loss of the adenine methylase in *dam⁻* strains affects the synchrony of origin firing, the mutant strains of bacteria grow normally, showing that there must be other ways of preventing rereplication. The most remarkable testimony to the robustness of the bacterial cell cycle is provided

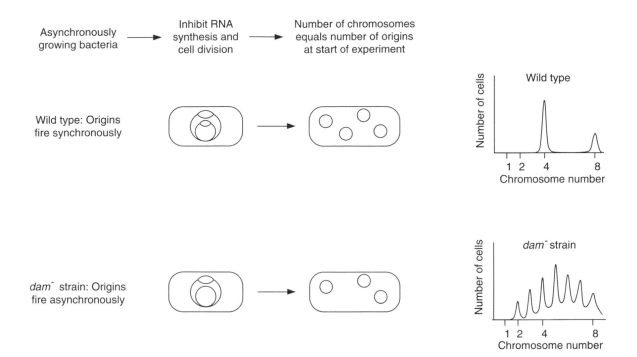

Asynchronously growing bacteria → Inhibit RNA synthesis and cell division → Number of chromosomes equals number of origins at start of experiment

Wild type: Origins fire synchronously

dam⁻ strain: Origins fire asynchronously

by mutant strains that allow cells to replicate even though they lack the normal replication origin and DnaA. Growth and chromosome replication are still well coordinated in these strains, showing that important undiscovered mechanisms exist to control DNA replication.

CHROMOSOME SEGREGATION AND SEPTATION

At the end of the cell cycle bacteria segregate their chromosomes and then divide by forming a septum that cuts the cell in two. Septation is due to a change in the way the cell wall grows; during elongation the wall grows as a cylinder that extends the rod-shaped cell, but during septation it curves inward to bisect the cell. We know a great deal about the synthesis of cell wall peptidoglycans because the enzymes that catalyze this process are targets for penicillin and related antibiotics. But although the enzymes responsible for the different modes of cell growth have been identified, the proteins that regulate their activity to initiate septation are less well understood.

FIGURE 11-8 DNA methylation is required for synchronous DNA replication. To determine the number of replication origins present in individual cells, a population of growing bacteria is treated with an inhibitor of RNA synthesis (which blocks any new initiation of DNA replication) and an inhibitor of cell division. Thus each origin present when the drugs are added gives rise to one chromosome. After all replication has finished, the population is subjected to flow cytometry, which measures the number of chromosomes in individual cells, generating the graphs on the right of the diagram. Reference: Boye, E., & Lobner-Oleson, A. *Cell* **62,** 981–989 (1990).

Periseptal annuli direct bacterial cell division

In eukaryotes chromosome segregation and cell division are organized by microtubules nucleated by microtubule organizing centers. Although bacteria lack microtubules, they do have specialized structures called **periseptal**

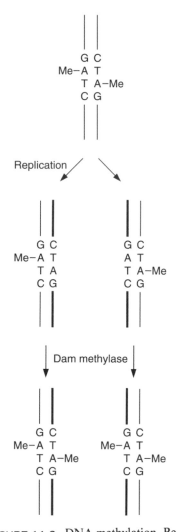

FIGURE 11-9 DNA methylation. Before DNA replication the adenines in the sequence GATC are methylated, but immediately after replication the newly synthesized strand (thicker line) is unmethylated. The Dam methylase methylates the newly synthesized strand after a lag that is short for most regions of the chromosome but long for the replication origin.

annuli that organize cell division. Periseptal annuli are a pair of closely spaced rings that run around the circumference of the cell, forming sites at which the inner and outer membranes communicate with each other (see Figure 11-1). At birth each cell has a pair of periseptal annuli located at its center and a single **polar annulus** at each end. Early in the cell cycle, two new pairs of periseptal annuli are formed, one on either side of the central pair. At first these new pairs are incomplete, occupying only a fraction of the circumference of the cell. As they migrate to positions midway between the central pair and the ends of the cell, they grow around the circumference of the cell to become complete annuli. At cell division the septum forms between the two members of the original central pair of annuli, which are thereby converted to the unpaired polar annuli found at the ends of the newborn cells (Figure 11-11).

It is has been speculated that chromosomes are attached to the membrane at the site of the periseptal annuli and that this attachment plays an important role in chromosome segregation. In such a model, the replicated chromosomes would detach from the central annuli late in the cell cycle and separate from each other. At some time between segregation and the beginning of the next cell cycle, the segregated chromosomes would become attached to the new pairs of periseptal annuli one-quarter and three-quarters along the length of the cell ensuring that the chromosomes would be located in the center of the newborn cells.

RELATIVE REPLICATION CAPACITY OF DIFFERENT FORMS OF PLASMID DNA

E. coli strain	Unmethylated	Hemimethylated	Methylated
Wild type	100	100	100
Methylase defective (*dam⁻*)	100	1	1

FIGURE 11-10 Hemimethylated DNA does not replicate. The effect of methylation on the replication of DNA plasmids that contain the *E. coli* replication origin was assayed by transfecting the plasmid in different methylation states into wild-type and methylation-deficient *(dam⁻)* cells. Only those cells that can replicate the plasmid many times give rise to colonies that express a gene on the plasmid. All three forms of DNA replicate in wild-type cells, but only unmethylated DNA replicates efficiently in *dam⁻* mutants, suggesting that initiation cannot occur on hemimethylated replication origins. Fully methylated DNA replicates only once in *dam⁻* cells because the first round of DNA replication produces two hemimethylated plasmid molecules. Reference: Russel, D. W., & Zinder, N. D. *Cell* **50**, 1071–1079 (1987).

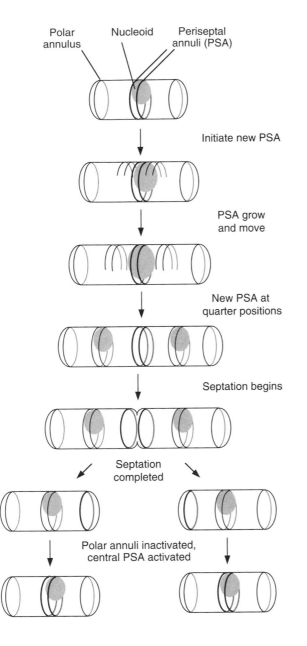

Polar annulus   Nucleoid   Periseptal annuli (PSA)

Initiate new PSA

PSA grow and move

New PSA at quarter positions

Septation begins

Septation completed

Polar annuli inactivated, central PSA activated

FIGURE 11-11 Periseptal annuli and septation. Cells are born with a single pair of periseptal annuli (PSA) that produce two new partial copies, which grow around the circumference of the cell as they migrate to positions one-quarter and three-quarters along the length of the cell. At septation the central periseptal annuli direct the formation of the septum. Septation converts the two rings of the central pairs of periseptal annuli to single polar annuli in the daughter cells and the periseptal annuli at the quarter positions to new central periseptal annuli. During or immediately after septation, the septation-inducing activity of the new polar annuli is inactivated.

Feedback controls regulate septation

How do bacterial cells ensure that chromosome replication is complete before cell division occurs? In eukaryotes the separation between the end of DNA replication and the onset of mitosis is clear-cut, and the presence of replicating DNA activates a feedback control that blocks entry into mitosis. In rapidly growing bacteria, however, DNA replication occurs throughout the cell cycle, and thus controls that responded to replicating DNA would arrest the normal cell cycle. Instead, bacteria appear to be able to use termination of a round of replication as a trigger for cell division.

Although termination of replication could induce cell division directly, bacterial cell division appears to be regulated by the ability of chromosomes at the center of the cell to inhibit septation. This feedback control was discovered by analyzing temperature-sensitive mutations in **DNA gyrase,** one of two type II DNA topoisomerases that are involved in resolving the topological linkage of the two replicated but intertwined daughter chromosomes. When certain temperature-sensitive gyrase mutants are shifted to the restrictive temperature, chromosome segregation is blocked, although cell growth and DNA replication continue normally. Shortly after the shift the cells have a single nucleoid and fail to divide, even after they have exceeded the threshold length for cell division. When cells reach twice the threshold length, the DNA is still at the center of the cell, and the central periseptal annuli are not used for septation. Instead, septation occurs at the periseptal annuli located at one-quarter and three quarters of the cell length to generate two cells that lack DNA, leaving a central cell that contains all the DNA (Figure 11-12).

We do not know how cells monitor the location of the nucleoids and use the information to control septation. The observation that some mutants that fail to segregate their chromosomes form septa that bisect the unsegregated chromosomes suggests that it is not merely the physical proximity of nucleoids that inhibits septation. There is an interesting but equally poorly understood parallel in eukaryotic cells, in which the presence of chromosomes close to the spindle poles can prevent cells from entering anaphase (see pages 148–149).

A second control that regulates septation is a feedback control activated by the presence of DNA damage. This pathway is perhaps the best understood of all feedback controls and is part of a complex response to DNA damage called the **SOS pathway.** The response is initiated by the RecA protein, which is best known for its ability to pair DNA molecules during recombination. Damaged DNA inhibits septation by inducing RecA to participate as a cofactor in the cleavage of a protein called LexA. Because intact LexA represses

FIGURE 11-12 Nearby DNA blocks septation. In a wild-type cell replicated nucleoids separate from each other at the termination of replication, and septation occurs at the central periseptal annuli. In a particular temperature-sensitive DNA gyrase mutant, the chromosome replicates normally, but daughter chromosomes cannot segregate from each other. At the time the wild-type cell divides, the mutant cell does not divide because the unsegregated nucleoid inhibits septation at the central periseptal annuli. By the time the cell has doubled in size again, however, the periseptal annuli at the quarter positions have matured and are used to produce daughter cells lacking DNA. Reference: Begg, K. J., & Donachie, W. D. *New Biol.* **3,** 475–486 (1991).

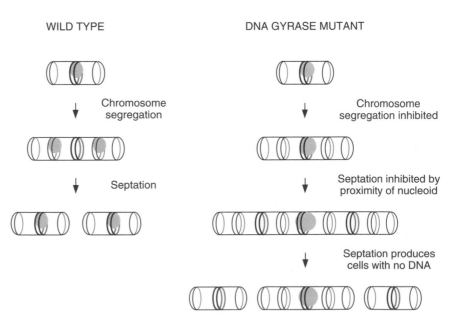

WILD TYPE   DNA GYRASE MUTANT

↓ Chromosome segregation   ↓ Chromosome segregation inhibited

↓ Septation   ↓ Septation inhibited by proximity of nucleoid

                              ↓ Septation produces cells with no DNA

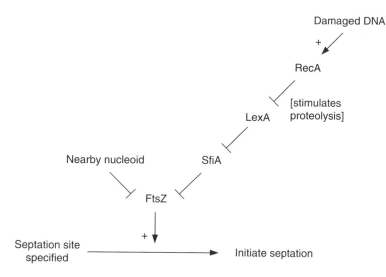

FIGURE 11-13 Feedback control of septation. Damaged DNA activates the RecA protein, which stimulates the destruction of LexA. The destruction of LexA allows the *sfiA* gene to be expressed, and SfiA in turn inhibits the ability of FtsZ to induce septation.

the transcription of the genes of the SOS pathway, its proteolysis induces the SOS response, including the expression of a protein called SfiA that inhibits septation (Figure 11-13).

FtsZ induces the septation of bacterial cells

The process of cell division has been studied by isolating a large collection of mutants that block cell division but not DNA replication. The *fts*^ts (filamentation temperature sensitive) mutants form long filaments with many nucleoids at the restrictive temperature. The key protein that induces septation appears to be the product of the *ftsZ*^+ gene. Monitoring the localization of the FtsZ protein during the cell cycle suggests that FtsZ plays an important role in initiating septation (Figure 11-14). Prior to septation FtsZ is distributed throughout the cell, but at the onset of septation it forms a ring at the site of the developing septum, and as the ring contracts FtsZ remains at its leading edge. The FtsZ protein is a GTPase encouraging the speculation that it forms part of a contractile ring that directs bacterial septation, much as the cortical actomyosin ring directs the cleavage of animal cells. The importance of FtsZ in septation is emphasized by the observation that the ability of SfiA to inhibit cell division can be overcome by overexpression of FtsZ.

Specific inhibitors restrict the site of septation to the central periseptal annuli

Theoretically, septation of an *E. coli* cell could occur at one of five possible sites: the pair of central periseptal annuli, the two new pairs of periseptal annuli at the one-quarter and three-quarter positions, and the two polar annuli. In practice, however, septation never occurs at the new periseptal annuli. Wild-type cells always septate at the central periseptal annuli, but *min*^− (minicell) mutants can septate at either the periseptal annuli or one of the polar annuli (Figure 11-15). Thus in wild-type cells the products of the

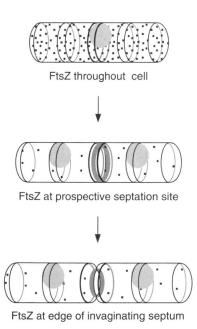

FtsZ throughout cell

FtsZ at prospective septation site

FtsZ at edge of invaginating septum

FIGURE 11-14 Localization of FtsZ at the site of septation. The FtsZ protein is required for septation, and its distribution during the cell cycle suggests that FtsZ plays a key role in initiating septation. Early in the cell cycle FtsZ is distributed throughout the cell, but just before septation FtsZ becomes localized in a ring at the future site of septation. Once septation begins, the FtsZ ring shrinks with the invaginating septum, as if it were a contractile structure driving septation. Reference: Bi, E., & Lutkenhaus, J. *Nature* **354**, 161–164 (1991).

FIGURE 11-15 Regulation of septation sites. In a wild-type cell (top left) the central periseptal annuli are always used for septation. Mutations in the *minE* gene block septation completely, while mutations in the *minC* or *minD* genes allow cells to use the polar annuli as well as the central periseptal annuli. Using the polar annuli produces miniature cells that lack DNA. Reference: De Boer, P., Cook, W. R., & Rothfield, L. I. *Ann. Rev. Genet.* **24,** 249–274 (1990).

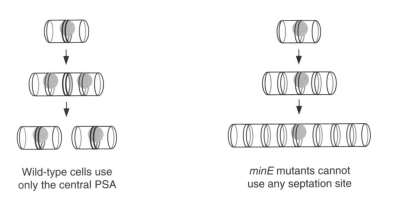

Wild-type cells use
only the central PSA

minE mutants cannot
use any septation site

minC and *minD* mutants can use central *or* polar PSA

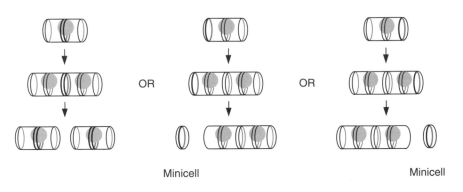

Minicell

Minicell

min⁺ genes must prevent septation at the polar annuli. Even in *min*⁻ mutants, however, only one septum is formed in each cell cycle, probably because cells produce only enough FtsZ to induce one septum: *min*⁻ mutants that also overexpress FtsZ can septate at more than one site simultaneously.

The *min* operon consists of three genes: *minC*⁺, *minD*⁺, and *minE*⁺. Inactivating *minC* or *minD* produces the minicell phenotype, while overexpressing either of these genes blocks all septation, suggesting that they encode functions that can inhibit septation. Inactivating *minE*⁺ blocks all septation, suggesting that this gene acts either to block the action of MinC and MinD at the central periseptal annuli or to localize their inhibitory activity to the polar annuli.

Bacterial chromosome segregation is poorly understood

If the mechanism of bacterial DNA replication is the best understood aspect of any cell cycle, bacterial chromosome segregation is among the most mysterious. Chromosome segregation appears to occur in two steps. During DNA replication the two nucleoids become visibly separate when viewed by electron microscopy (presumably as a result of topoisomerase-mediated decatenation), although they remain so close to each other that the light microscope sees them as a single object. Once replication has finished and the cell has reached the threshold length, the chromosomes segregate rapidly to their new positions.

Two types of mutation affect bacterial chromosome segregation. The first is mutations in the DNA topoisomerases whose activity is essential to allow the daughter chromosomes to separate from each other topologically. The second type is mutations that do not completely block chromosome segregation but increase the fraction of cells that fail to receive a chromosome when they divide from about 0.03% to more than 10%. These mutants were found by an ingenious genetic screen that identifies strains that produce a high fraction of cells without chromosomes. This screen has identified mutations in a previously known outer membrane protein and also in $mukB^+$, a novel gene. The $mukB^+$ gene encodes a large protein that has a long region that could form an elongated, rigid tail, as well as an amino-terminal region that shows homology to the eukaryotic dynamins. The dynamins are microtubule-binding proteins that play important roles in the intracellular transport of organelles, making MukB an interesting candidate for a bacterial protein that is directly involved in chromosome movement.

CONCLUSION

Our scanty knowledge of the molecules that regulate the bacterial cell cycle makes it impossible to accurately assess the similarity between eukaryotic and prokaryotic cell cycles. The initiation of DNA replication is the only cell cycle process that has been well characterized in both types of organism, and the basic principles that establish replication forks appear to be conserved. This similarity encourages us to hope that other aspects of the bacterial and eukaryotic cell cycles have been equally well conserved so that advances in our understanding of one will aid research in the other. Even if this hope is not fulfilled, the wealth of questions that our new understanding of the cell cycle has triggered promises to make studying the cell cycle during the 1990s just as exciting as it was in the 1980s.

SELECTED READINGS

General

Neidhart, F. C., ed. *Escherichia coli and Salmonella typhimurium: Cellular and Molecular Biology* (American Society for Microbiology, Washington, D.C., 1987). *A detailed review of all aspects of the cell biology of the two most-studied prokaryotes.*

Kornberg, A., & Baker, T. A. *DNA Replication* (W. H. Freeman, New York, 1992). *A detailed, up-to-date, and comprehensive review of DNA replication in eukaryotes and prokaryotes.*

Cooper, S. *Bacterial Growth and Division.* (Academic Press, San Diego, 1991). *A careful review of the physiological analysis of cell growth, DNA replication, and cell division in bacteria.*

Reviews

De Boer, P., Cook, W. R., & Rothfield, L. I. Bacterial cell division. *Ann. Rev. Genet.* **24,** 249–274 (1990). *An excellent description of the behavior of periseptal annuli and mutants that affect cell division.*

Marians, K. J. Prokaryotic DNA replication. *Ann. Rev. Biochem.* **61,** 673–719 (1992). *A detailed review of the control of DNA replication in prokaryotic cells and viruses.*

Hiraga, S. Chromosome and plasmid partition in *Escherichia coli*. *Ann. Rev. Biochem.* **61**, 283–306 (1992). *A comprehensive review of what we know about chromosome segregation in* E. coli.

Funnell, B. E. Participation of the bacterial membrane in DNA replication and chromosome partition. *Trends Cell Biol.* **3**, 20–25 (1993). *A concise review of the evidence that association with the cell membrane controls bacterial DNA replication.*

Donachie, W. D. The cell cycle of *Escherichia coli*. *Ann. Rev. Microbiol.* **47**, in press (1993).

Original Articles

Cooper, S., & Helmstetter, C. E. Chromosome replication and the division cycle of *Escherichia coli* B/r. *J. Mol. Biol.* **1**, 519–540 (1968). *The original demonstration that it takes* E. coli *cells longer to replicate their DNA than it takes them to divide.*

Donachie, W. D., & Begg, K. J. Cell length, nucleoid separation, and cell division of rod-shaped and spherical cells of *Escherichia coli*. *J. Bacteriol.* **171**, 4633–4639 (1989). *Using* E. coli *mutants that alter cell shape to demonstrate that cells have to reach a critical length before they can segregate their chromosomes.*

Boye, E., & Lobner-Oleson, A. The role of dam methyltransferase in the control of DNA replication in *E. coli*. *Cell* **62**, 981–989 (1990). *Altering the extent of DNA methylation affects the timing of replication initiation.*

Bi, E., & Lutkenhaus, J. FtsZ ring structure associated with division in *Escherichia coli*. *Nature* **354**, 161–164 (1991). *The behavior of the FtsZ protein suggests that it may assemble into a structure that guides septation.*

Niki, H., Jaffe, A., Imamura, R., Ogura, T., & Hiraga, S. The new gene mukB codes for a 177 kd protein with coiled-coil domains involved in chromosome partitioning of *E. coli*. *EMBO J.* **10**, 183–193 (1991). *The analysis of a mutation that disrupts bacterial chromosome segregation.*

Appendix
Genes That Affect the Cell Cycle

An alphabetical index of genes that affect the cell cycle appears on pages 240–243.

Entry	Gene	Description	Reference	Genbank Name

Budding Yeast

CDC Genes

Entry	Gene	Description	Reference	Genbank Name
1	*CDC2*	Also called *POL3*. Encodes the catalytic subunit of DNA polymerase δ (also called DNA polymerase III). *ts* mutants arrest in late S phase.	Budd, M.E. & Campbell, J.L. *Mol Cell Biol* **13**, 496–505 (1993).	YSCCDC2
2	*CDC3*	With *CDC10, CDC11,* and *CDC12,* forms a family of related genes whose products are needed for assembling a ring of 10-nm filaments at the neck between mother and daughter cells. ts mutations in any of the genes prevent cytokinesis.	Kim, H.B., Haarer, B.K. & Pringle, J.R. *J Cell Biol* **112**, 535–544 (1991).	
3	*CDC4*	Encodes a protein with homology to certain β subunits of trimeric G proteins. *ts* mutants block initiation of DNA synthesis and spindle pole body separation but not budding. Interacts genetically with *CDC34* and *CDC53*. Suppressed by overexpression of *CLB2*.	Smith, S.A., Kumar, P., Johnston, I. & Rosamond, J. *Mol Gen Genet* **235**, 285–291 (1992).	YSCCDC4G
4	*CDC5*	Also called *PKX2*. Encodes a protein kinase. *ts* mutants arrest after nuclear division, but before cytokinesis. Little studied.	Sharon, G. & Simchen, G. *Genetics* **125**, 475–485 (1990).	YSCPKX2A

Entry	Gene	Description	Reference	Genbank Name
5	CDC6	Cloned but not homologous to known genes. *ts* mutants arrest cells late in S phase. Circumstantial evidence for a role in the initiation of DNA synthesis. Normally transcribed only during late G1 and S phase. Constitutive expression delays mitosis.	Bueno, A. & Russell, P. *Embo J* **11**, 2167–2176 (1992).	YSCCDC6G
6	CDC7	Encodes a protein kinase required for the initiation of DNA synthesis in mitotic but not meiotic cell cycles (although it is required later in the meiotic cell cycle). *ts* mutants fail to initiate DNA synthesis, and some alleles can be suppressed by overexpression of *DBF4*.	Hollingsworth, R.J. & Sclafani, R.A. *Proc Natl Acad Sci U S A* **87**, 6272–6276 (1990).	YSCCDC7
7	CDC8	Encodes thymidylate kinase which converts dTMP into dTDP. *ts* mutants rapidly cease DNA synthesis when shifted to 37 °C.	Jiang, Z.R., *et al. J Biol Chem* **266**, 18287–18293 (1991).	YSCCDC8
8	CDC9	Encodes DNA ligase. *ts* mutants arrest in G2 with single-stranded nicks and gaps in DNA; arrest overcome by inactivation of the *RAD9*-dependent feedback control pathway.	Tomkinson, A.E., Tappe, N.J. & Friedberg, E.C. *Biochemistry* **31**, 11762–11171 (1992).	YSCCDC9
9	CDC10	See *CDC3*.	Ford, S.K. & Pringle, J.R. *Dev Genet* **12**, 281–292 (1991).	
10	CDC11	See *CDC3*.	Ford, S.K. & Pringle, J.R. *Dev Genet* **12**, 281–292 (1991).	
11	CDC12	See *CDC3*.	Kim, H.B., Haarer, B.K. & Pringle, J.R. *J Cell Biol* **112**, 535–544 (1991).	
12	CDC13	Cloned but not homologous to known genes. *ts* mutants arrest in G2 of the mitotic and meiotic cell cycles with damaged DNA; arrest overcome by inactivation of *RAD9*-dependent feedback control pathway.	Weber, L. & Byers, B. *Genetics* **131**, 55–63 (1992).	YSCCDC13
13	CDC14	Encodes a protein homologous to tyrosine phosphatases. *ts* mutants arrest in mitosis with two separated masses of DNA (probably anaphase). Circumstantial evidence for roles in initiation of DNA synthesis and control of gene expression suggests a function in chromatin structure.	Wan, J., Xu, H. & Grunstein, M. *J Biol Chem* **267**, 11274–11280 (1992).	YSCCDC14A

Entry	Gene	Description	Reference	Genbank Name
14	*CDC15*	Encodes a protein homologous to protein kinases. *ts* mutants arrest in mitosis with a long spindle and two separated masses of DNA and can be suppressed by overexpression of *SPO12*.	Schweitzer, B. & Philippsen, P. *Yeast* **7**, 265–273 (1991).	YSCCDC15
15	*CDC16*	*CDC16* and *CDC23* both encode proteins containing multiple tandem copies of a loosely conserved 34-amino-acid repeat found in several other genes implicated in nuclear events. *ts* mutants in either gene arrest in mitosis with a short spindle and unseparated chromosomes. Arrest independent of *MAD/BUB*-dependent feedback control. Interacts genetically with *CDC23*.	Sikorski, R.S., Boguski, M.S., Goebl, M. & Hieter, P. *Cell* **60**, 307–317 (1990).	YSCCDC16
16	*CDC17*	Also called *POL1*. Encodes the catalytic subunit of DNA polymerase α (also called polymerase I). *ts* mutants arrest in mid to late S phase.	Barker, D.G., White, J.H. & Johnston, L.H. *Eur J Biochem* **162**, 659–667 (1987).	YSPPOL1G
17	*CDC18*	Not cloned. Leaky *ts* mutants with mutant phenotype similar to *cdc5^ts*. Little studied.	Hartwell, L.H., Mortimer, R.K., Culotti, J. & Culotti, M. *Genetics* **74**, 267–286 (1973).	
18	*CDC19*	Also called *PYK1*. Encodes pyruvate kinase. *ts* mutants arrest prior to Start and do not increase in mass.	Moore, P.A., Bettany, A.J. & Brown, A.J. *J Gen Microbiol* **136**, 2359–2366 (1990).	YSCPYK1
19	*CDC20*	Encodes a protein with homology to β subunits of trimeric G proteins. *ts* mutants arrest in mitosis with nuclear microtubules splayed out rather than arranged in a discrete spindle.	Sethi, N., *et al. Mol Cell Biol* **11**, 5592–5602 (1991).	YSCDC20
20	*CDC21*	Also called *TMP1*. Encodes thymidylate synthetase which converts dUMP to dTMP. *ts* mutants rapidly cease DNA synthesis.	Poon, P.P. & Storms, R.K. *J Biol Chem* **266**, 16808–16812 (1991).	YSCTMP1
21	*CDC23*	See *CDC16*.	Sikorski, R.S., Boguski, M.S., Goebl, M. & Hieter, P. *Cell* **60**, 307–317 (1990).	YSCCDC23X

Entry	Gene	Description	Reference	Genbank Name
22	CDC24	Also called *CLS4*. Encodes a putative calcium-binding protein. A human homolog catalyzes guanine nucleotide exchange on the human homolog of Cdc42. *ts* mutants fail to bud, but other cell cycle processes continue normally. Interacts genetically with *CDC42* and genes involved in bud site selection.	Bender, A. & Pringle, J.R. *Yeast* **8**, 315–323 (1992).	YSCCLS4A
23	CDC25	Encodes a guanine nucleotide exchange protein that acts on Ras. *ts* mutants arrest prior to Start and fail to increase in mass.	Gross, E., Goldberg, D. & Levitzki, A. *Nature* **360**, 762–765 (1992).	YSCCDC25G
24	CDC28	Encodes a protein kinase that associates with cyclins. Activity essential for Start and mitosis, but *ts* mutants arrest predominantly at Start. Homologs (Cdc2 in fission yeast) in all eukaryotes. Not required for premeiotic DNA synthesis but is required for entry into meiosis I.	Sorger, P.K. & Murray, A.W. *Nature* **355**, 365–368 (1992).	YSCCDC28
25	CDC30	Encodes one of two forms of phosphoglucose isomerase. *ts* mutants arrest late in mitosis, but the arrest is leaky and suppressed by nonfermentable carbon sources.	Dickinson, J.R. *J Gen Microbiol* **137**, 765–770 (1991).	
26	CDC31	Encodes a small calcium-binding protein whose activity is required for spindle pole body duplication.	Winey, M., Goetsch, L., Baum, P. & Byers, B. *J Cell Biol* **114**, 745–754 (1991).	YSCCDC31
27	CDC32	Not cloned. *ts* mutants arrest as unbudded G1 cells. One *cdc32^ts* mutation supresses *cdc51^ts*. Little studied.	Moir, D., Stewart, S.E., Osmond, B.C. & Botstein, D. *Genetics* **100**, 547–563 (1982).	
28	CDC33	Encodes eIF-4E (eukaryotic initiation factor 4E), which binds to the cap at the 5′ end of mRNA molecules. *ts* mutants arrest prior to Start and fail to increase in mass. Δ*bcy1 cdc33^ts* cells arrest at random points in the cycle.	Brenner, C., *et al. Mol Cell Biol* **8**, 3556–3559 (1988).	YSCCDC33
29	CDC34	Encodes a ubiquitin-conjugating enzyme, which appears to be involved in the degradation of Cln1 and Cln2. *ts* mutants block initiation of DNA synthesis and spindle pole body separation but not budding. Interacts genetically with *CDC4* and *CDC53*.	Kolman, C.J., Toth, J. & Gonda, D.K. *Embo J* **11**, 3081–3090 (1992).	YSCUBC3

Entry	Gene	Description	Reference	Genbank Name
30	CDC35	Also called *CYR1*. Encodes adenylate cyclase. *ts* mutants arrest prior to Start and fail to increase in mass. May interact physically with Ira1. Δ*cdc35* cells can proliferate, albeit very slowly.	Mitts, M.R., Bradshaw, R.J. & Heideman, W. *Mol Cell Biol* **11**, 4591–4598 (1991).	YSCCYR1
31	CDC36	*CDC36* and *CDC39* have both been cloned, but neither is homologous to known genes. *cdc36*[ts] and *cdc39*[ts] mutants have two defects: activation of mating factor signaling pathway (which leads to arrest at Start) and a secondary defect that leads to a slower arrest at random points in the cell cycle.	Neiman, A.M., Chang, F., Komachi, K. & Hersko-witz, I. *Cell Regul* **1**, 391–401 (1990).	YSCCDC36
32	CDC37	Cloned but not homologous to known genes. Mutant arrests with similar phenotype to *cdc28*[ts].	Ferguson, J., Ho, J.Y., Peter-son, T.A. & Reed, S.I. *Nucleic Acids Res* **14**, 6681–6697 (1986).	YSCCDC37
33	CDC39	See *CDC36*.		YSCCDC39
34	CDC40	Cloned but not sequenced. *ts* mutants arrest with partially replicated DNA. Defective in DNA repair at 37°C.	Kassir, Y., Kupiec, M., Sha-lom, A. & Simchen, G. *Curr Genet* **9**, 253–257 (1985).	
35	CDC42	Encodes a small G protein. *ts* mutants fail to bud, but other cell cycle pro-cesses continue normally. Interacts genetically with *CDC24*, *CDC43*, and genes involved in bud site selec-tion.	Johnson, D.I. & Pringle, J.R. *J Cell Biol* **111**, 143–152 (1990).	YSCCDC42
36	CDC43	Encodes the β subunit of a prenyl-transferase, which modifies Cdc42. *ts* mutants fail to bud, but other cell cycle processes continue normally. Interacts genetically with *CDC42* and genes involved in bud site selec-tion.	Finegold, A.A., *et al. Proc Natl Acad Sci U S A* **88**, 4448–4452 (1991).	YSCCDC43A
37	CDC44	Not cloned. A single *cs* mutant arrests in G2 with large budded cells. Little studied.	Moir, D. & Botstein, D. *Ge-netics* **100**, 565–577 (1982).	
38	CDC45	Not cloned. A *cs* mutant is unable to initiate DNA synthesis. Interacts ge-netically with *CDC46* and *CDC47*.	Hennessy, K.M., Lee, A., Chen, E. & Botstein, D. *Genes Dev* **5**, 958–969 (1991).	

Entry	Gene	Description	Reference	Genbank Name
39	CDC46	Also called *MCM5*. Encodes a protein with homology to other proteins thought to be involved in initiation of DNA replication (budding yeast Mcm2 and Mcm3, and fission yeast Cdc21 and Nda4). ts mutants cannot initiate DNA replication. Cdc46 enters the nucleus at anaphase, making it a candidate for licensing factor.	Chen, Y., Hennessy, K.M., Botstein, D. & Tye, B.K. *Proc Natl Acad Sci U S A* **89,** 10459–10463 (1992).	
40	CDC47	Not cloned. Mutant phenotype similar to *cdc46*^{ts}.	Hennessy, K.M., Lee, A., Chen, E. & Botstein, D. *Genes Dev* **5,** 958–969 (1991).	
41	CDC48	Encodes a protein with strong homology to oligomeric ATPases involved in membrane fusion and organelle biogenisis. *ts* mutants arrest in G2 with a single spindle pole body and aberrantly distributed microtubules.	Frohlich, K.U., *et al. J Cell Biol* **114,** 443–453 (1991).	YSCCDC48
42	CDC49	Not cloned. Several *cdc49*^{cs} mutants and a single *cdc50*^{cs} mutant arrest with a small bud and a single nucleus. Little studied.	Moir, D., Stewart, S.E., Osmond, B.C. & Botstein, D. *Genetics* **100,** 547–563 (1982).	
43	CDC50	Not cloned. See *CDC49*.		
44	CDC51	Not cloned. A *ts* mutant arrests with a large bud and undivided nucleus and is suppressed by *cdc32*^{ts}. Little studied.	Moir, D., Stewart, S.E., Osmond, B.C. & Botstein, D. *Genetics* **100,** 547–563 (1982).	
45	CDC53	Not cloned. *cdc53*^{ts} mutants have a similar phenotype to *cdc4*^{ts} and *cdc34*^{ts}.		
46	CDC54	Not cloned. *cs* mutants are unable to initiate DNA synthesis. Interacts genetically with *CDC46* and *CDC47*.	Hennessy, K.M., Lee, A., Chen, E. & Botstein, D. *Genes Dev* **5,** 958–969 (1991).	
47	CDC55	Encodes a protein homologous to the β subunit of protein phosphatase 2A. Δ*cdc55* cells are *cs* for cytokinesis. A *cs* mutant is suppressed by *bem2*⁻.	Healy, A.M., *et al. Mol Cell Biol* **11,** 5767–5780 (1991).	YSCCDC55G
48	CDC60	Encodes leucyl-tRNA synthetase. Arrests prior to Start, probably due to decreased rates of protein synthesis.	Hohmann, S. & Thevelein, J.M. *Gene* **120,** 43–49 (1992).	YSCCDC60G
49	CDC63	Also called *PRT1*. Cloned but not homologous to known genes. *ts* mutants are defective in the initiation of protein synthesis and cannot pass Start.	Hanic, J.P., Johnston, G.C. & Singer, R.A. *Exp Cell Res* **172,** 134–145 (1987).	YSCPRT1

Entry	Gene	Description	Reference	Genbank Name
50	CDC65	Also called DNA33. Not cloned. cdc65[ts] and cdc67[ts] mutants arrest cells prior to Start but continue to increase in mass.	Prendergast, J.A., et al. Genetics 124, 81–90 (1990).	
51	CDC67	Not cloned. See CDC65.		
52	CDC68	Also called SPT16. Encodes a protein that is probably involved in control of transcription. ts mutants arrest prior to Start. This arrest is overcome by overexpression of CLN2.	Rowley, A., Singer, R.A. & Johnston, G.C. Mol Cell Biol 11, 5718–5726 (1991).	YSCCDC68
53	CDC72	Also called NMT1. Encodes N-myristol transferase, the enzyme that attaches myristol groups to proteins thus targeting them to membrane locations. ts mutants fail to myristolate Scg1, leading to constitutive activation of the mating factor signaling pathway.	Stone, D.E., et al. Genes Dev 5, 1969–1981 (1991).	YSCNMT
54	CDC73	Not cloned. ts mutants constitutively activate the mating-factor-response pathway.	Reed, S.I., Ferguson, J. & Jahng, K.Y. Cold Spring Harb Symp Quant Biol 2, 621–627 (1988).	

Regulators of Cdc28

Entry	Gene	Description	Reference	Genbank Name
55	CLN1	CLN1, CLN2, and CLN3 encode functionally redundant G1 cyclins. CLN1 and CLN2 are closely related to each other and much more distantly related to CLN3. Cln1 and Cln2 mRNA and protein levels fluctuate strongly throughout the cell cycle, reaching maximal levels at Start, whereas those for Cln3 show little or no variation. Δcln1 Δcln2 Δcln3 triple mutants cannot pass Start, but all double deletion strains are viable.	Cross, F.R. & Tinkelenberg, A.H. Cell 65, 875–883 (1991).	YSCCLN1A
56	CLN2	See CLN1.		YSCCLN2A
57	CLN3	Also called DAF1 and WHI1. Encodes a G1 cyclin. CLN3[D] mutants, which produce a truncated stable protein, pass Start at a smaller size than wild-type. See CLN1.	Tyers, M., Tokiwa, G., Nash, R. & Futcher, B. Embo J 11, 1773–1784 (1992).	YSCWHI1
58	HCS26	Encodes a putative G1 cyclin. Interacts genetically with SWI4 but not with CLN genes.	Ogas, J., Andrews, B.J. & Herskowitz, I. Cell 66, 1015–1026 (1991).	YSCG1CYC

Entry	Gene	Description	Reference	Genbank Name
59	*CLB1*	*CLB1*, *CLB2*, *CLB3*, and *CLB4* encode partially redundant mitotic (B-type) cyclins. *CLB1* and *CLB2* are closely related to each other, as are *CLB3* and *CLB4*. Cells can survive with only *CLB2*. Mutants that lack all four of this group of cyclins proceed through Start, bud, and synthesize DNA but fail to assemble a spindle.	Fitch, I., *et al. Mol Biol Cell* **3**, 805–818 (1992).	YSCCLB1
60	*CLB2*	See *CLB1*.		YSCCLB2
61	*CLB3*	See *CLB1*.		YSCMCYBA
62	*CLB4*	See *CLB1*.		YSCMCYBB
63	*CLB5*	*CLB5* and *CLB6* encode redundant B-type cyclins that are closely related to each other. Their primary function appears to be inducing DNA synthesis. *CLB5* and *CLB6* show partial redundancy with both the *CLN* genes and with *CLB1* to *CLB4*.	Epstein, C.B. & Cross, F.R. *Genes Dev* **6**, 1695–1706 (1992).	YSCCLB5A
64	*CLB6*	See *CLB5*.	K. Nasmyth, personal communication	
65	*MIH1*	Encodes a homolog of fission yeast Cdc25. Δ*mih1* cells exhibit a G2 delay, which is converted into a G2 arrest by overexpressing fission yeast *wee1*.	Bueno, A. & Russell, P. *Embo J* **11**, 2167–2176 (1992).	J04846
66	*SWE1*	Encodes a homolog of fission yeast Wee1. Δ*swe1* suppresses the G2 delay seen in Δ*mih1* strains.	R. Booher, personal communication	
67	*CKS1*	Encodes a protein that binds Cdc28 and is homologous to fission yeast Suc1.	Hadwiger, J.A., Wittenberg, C., Mendenhall, M.D. & Reed, S.I. *Mol Cell Biol* **9**, 2034–2041 (1989).	YSCPKCK51
68	*SWI4*	Encodes the DNA-binding subunit of the Swi4/Swi6 transcription factor that induces the expression of *ClN1*, *ClN2*, and *HO* as cells approach Start.	Nasmyth, K. & Dirick, L. *Cell* **66**, 995–1013 (1991).	YSCSW14
69	*SWI6*	Encodes a transcriptional activator that associates with Swi4 to stimulate transcription of G1 cyclins and with Mpb1 to stimulate transcription of genes involved in DNA synthesis.	Lowndes, N.F., Johnson, A.L., Breeden, L. & Johnston, L.H. *Nature* **357**, 505–508 (1992).	YSCSW16
70	*MPB1*	Encodes the DNA-binding subunit of the Mpb1/Swi6 transcription factor that induces the expression of Clb5 and Clb6, as well as numerous genes involved in DNA replication.	K. Nasmyth, personal communication	

Entry	Gene	Description	Reference	Genbank Name

Cyclic AMP and growth control

Entry	Gene	Description	Reference	Genbank Name
71	TPK1	*TPK1, TPK2,* and *TPK3* encode three closely related catalytic subunits of cyclic AMP-dependent protein kinase. Simultaneous deletion of all three genes is lethal.	Toda, T., *et al. Cell* **50**, 277–287 (1987).	YSCTPKA
72	TPK2	See *TPK1.*		YSCTPKB
73	TPK3	See *TPK1.*		YSCTPKC
74	BCY1	Encodes the regulatory subunit of cyclic AMP-dependent protein kinase. Δ*bcy1* cells cannot sporulate or enter G0 and are sensitive to heat shock.	Werner, W.M., Brown, D. & Braun, E. *J Biol Chem* **266**, 19704–19709 (1991).	YSCCAPK
75	PDE1	Encodes the low-affinity cyclic AMP phosphodiesterase. Overexpression of *PDE1* or *PDE2* suppresses the effects of dominant, activating *RAS* mutants. Deletion of *PDE1* and *PDE2* mimics dominant, activating *RAS* mutants.	Nikawa, J., Sass, P. & Wigler, M. *Mol Cell Biol* **7**, 3629–3636 (1987).	YSCPPR
76	PDE2	Encodes the high-affinity cyclic AMP phosphodiesterase. See *PDE1.*	Sass, P., *et al. Proc Natl Acad Sci U S A* **83**, 9303–9307 (1986).	YSCPDE2
77	CYR1	Also called *CDC35* (30).		
78	CAP	Encodes a protein that associates with and regulates the activity of adenylate cyclase. The amino-terminal region of Cap is required for normal responses to Ras. The carboxyl-terminal region is required for normal cell morphology, budding pattern, actin morphology, and response to nutrient arrest.	Vojtek, A., *et al. Cell* **66**, 497–505 (1991).	YSCCAP
79	PFY1	Encodes profilin, a protein that binds both actin and phosphoinositides. Deletion mimics the phenotype of loss of the carboxyl-terminal region of Cap.	Vojtek, A., *et al. Cell* **66**, 497–505 (1991).	YSCPROFR
80	RAS1	*RAS1* and *RAS2* encode two related Ras proteins. Neither gene is essential, but deletion of both is lethal. Activated alleles prevent sporulation or entry into G0.	Gerst, J.E., Rodgers, L., Riggs, M. & Wigler, M. *Proc Natl Acad Sci U S A* **89**, 4338–4342 (1992).	YSCRAS1

Entry	Gene	Description	Reference	Genbank Name
81	*RAS2*	See *RAS1*.		YSCRAS2
82	*SCD25*	Encodes a guanine–nucleotide-exchange protein, whose C-terminal domain can suppress *cdc25^{ts}* mutants and catalyze GTP exchange on Ras. Overexpressing the full-length protein does not suppress *cdc25^{ts}*. Not essential.	Mistou, M.Y., *et al. Embo J* **11**, 2391–2397 (1992).	YSCSCD25
83	*IRA1*	*IRA1* and *IRA2* encode GTPase-activating proteins that stimulate GTP hydrolysis by Ras. Deletion of *IRA1* or *IRA2* suppresses *cdc25^{ts}*, prevents sporulation and entry into G0, and makes cells sensitive to heat shock	Tanaka, K., *et al. Cell* **60**, 803–807 (1990).	M24378
84	*IRA2*	See *IRA1*.		YSCIRA2A
85	*RAM1*	*RAM1* encodes the α subunit and *RAM2* the β subunit of the prenyl transferase that prenylates Ras1, Ras2, and **a** factor. Essential.	He, B., *et al. Proc Natl Acad Sci U S A* **88**, 11373–11377 (1991).	YSCDPR
86	*RAM2*	See *RAM1*. Also called *STE16, DPR1, SCG2*. Not essential.		YSCRAM2
87	*SNC1*	Encodes a homolog of proteins associated with synaptic vesicles in mammalian cells. Overexpression supresses deletion of the N terminus of Cap in cells that also contain activated Ras.	Gerst, J.E., Rodgers, L., Riggs, M. & Wigler, M. *Proc Natl Acad Sci U S A* **89**, 4338–4342 (1992).	YSCSP1DER
88	*MSI1*	Encodes a protein with homology to the β subunit of trimeric G proteins. Overexpression suppresses the phenotype of $\Delta ira1$ or activating Ras mutations.	Ruggieri, R., *et al. Proc Natl Acad Sci U S A* **86**, 8778–8782 (1989).	YSCMSI1
89	*RPI1*	Cloned but not homologous to known genes. Genetic evidence suggests that it stimulates activity of Ira1 and Ira2.	Kim, J.H. & Powers, S. *Mol Cell Biol* **11**, 3894–3904 (1991).	YSCRPI1
90	*SCH9*	Encodes a protein kinase. Overexpression allows cells lacking Cdc25, Ras, adenyl cyclase or cyclic AMP-dependent kinase to proliferate. $\Delta sch9$ cells are viable but proliferate slowly and have a prolonged G1 phase.	Denis, C.L. & Audino, D.C. *Mol Gen Genet* **229**, 395–399 (1991).	YSCSCH9
91	*YAK1*	Encodes a protein kinase whose activity is greatly increased by cell cycle arrest. *yak1⁻* mutants allow cells lacking Ras or cyclic AMP-dependent kinase to proliferate. Not essential.	Garrett, S., Menold, M.M. & Broach, J.R. *Mol Cell Biol* **11**, 4045–4052 (1991).	YSCYAK1

Entry	Gene	Description	Reference	Genbank Name

Kinases and phosphatases

Entry	Gene	Description	Reference	Genbank Name
92	PTP1	PTP1 and PTP2 encode related tyrosine phosphatases.	Guan, K.L., Deschenes, R.J., Qiu, H. & Dixon, J.E. *J Biol Chem* **266**, 12964–12970 (1991).	YSCPTPASE
93	PTP2	Two genes have been named PTP2. One encodes a close relative of PTP1 and cells can survive in the absence of both genes. The other is described below.		YSCRET1A
94	PTP2	Encodes a putative tyrosine phosphatase. Δptp2 cells are viable but heat sensitive.	Ota, I.M. & Varshavsky, A. *Proc Natl Acad Sci U S A* **89**, 2355–2359 (1992).	YSCPTP2A
95	DIS2	Also called GLC7. Encodes protein phosphatase 1. Essential for cell proliferation and probably acts on multiple substrates. Dis2 probably removes inhibitory phosphates from factors involved in the initiation of protein synthesis.	Wek, R.C., Cannon, J.F., Dever, T.E. & Hinnebusch, A.G. *Mol Cell Biol* **12**, 5700–5710 (1992).	YSCPP1A
96	GAC1	Encodes a protein with homology to mammalian regulators of phosphatase I. Deletion leads to reduced glycogen levels.	Francois, J.M., *et al. Embo J* **11**, 87–96 (1992).	YSCGAC1
97	PPH21	PPH21, PPH22, and PPH3 encode three protein phosphatase 2A catalytic sub-units. Deletion of all three genes is lethal, whereas the simultaneous deletion of PPH21 and PPH22 is lethal in some strain backgrounds but not others. Overexpression of PPH21 induces abnormal cell morphology	Sneddon, A.A., Cohen, P.T. & Stark, M.J. *Embo J* **9**, 4339–4346 (1990).	YSCPPH1G
98	PPH22	See PPH21.		YSCPPH22G
99	PPH3	See PPH21.	Ronne, H., Carlberg, M., Hu, G.Z. & Nehlin, J.O. *Mol Cell Biol* **11**, 4876–4884 (1991).	SCPPH3
100	TPD3	Encodes the regulatory α subunit of protein phosphatase 2A. Deletions have cs defects in cytokinesis. See CDC55 (47).	van Zyl, W., *et al. Mol Cell Biol* **12**, 4946–59 (1992).	YSCTPD3X
101	SIT4	Encodes a relative of protein phosphatase 2A that is required for CLN1 and CLN2 transcription and budding.	Sutton, A., Immanuel, D. & Arndt, K.T. *Mol Cell Biol* **11**, 2133–2148 (1991).	M24395

Entry	Gene	Description	Reference	Genbank Name
102	SSD1	Encodes a protein with homology to fission yeast Dis3. Not essential, but the phenotypes of Δsit4 and Δbcy1 are suppressed by some alleles of SSD1.	Sutton, A., Immanuel, D. & Arndt, K.T. Mol Cell Biol 11, 2133–2148 (1991).	YSCSIT4B
103	PPZ1	PPZ1 and PPZ2 encode closely related novel phosphatases that are homologous to both protein phosphatases 1 and 2A.	Da Cruz, E., et al. Biochim Biophys Acta 1089, 269–272 (1991).	RABPP2BWS
104	PPZ2	See PPZ1.		YSCPHOSPAZ
105	CNA1	Also called CMP1. CNA1 and CNA2 encode related protein phosphatase 2B catalytic subunits. Deletion of both genes does not kill cells but increases sensitivity to mating factors and slows recovery from exposure to mating factors.	Cyert, M.S., Kunisawa, R., Kaim, D. & Thorner, J. Proc Natl Acad Sci U S A 88, 7376–7380 (1991).	YSCCALA1
106	CNA2	Also called CMP2. See CNA1.		YSCCALA2
107	CNB1	Encodes the regulatory subunit of protein phosphatase 2B. Deletion has a similar phenotype to Δcna1 Δcna2.	Cyert, M.S. & Thorner, J. Mol Cell Biol 12, 3460–3469 (1992).	YSCCNB1
108	CMD1	Encodes yeast calmodulin. Essential. Mutants that cannot bind calcium are fully viable. A ts mutant buds slowly and dies as it goes through mitosis.	Davis, T.N. J Cell Biol 118, 607–617 (1992).	YSCCMD1
109	CMK1	CMK1 and CKM2 both encode calcium-calmodulin-dependent protein kinase. Strains that lack both genes are viable.	Ohya, Y., et al. J Biol Chem 266, 12784–12794 (1991).	YSCCMK1
110	CMK2	See CMK1.		YSCCMK2II
111	PKC1	Encodes protein kinase C. Δpkc1 cells lyse unless grown in medium of high osmotic strength.	Paravicini, G., et al. Mol Cell Biol 12, 4896–4905 (1992).	YSCPKC1A
112	YPK1	Also called MCK1. Encodes a tyrosine kinase whose deletion impairs centromere function, induction of meiosis, and spore formation.	Neigeborn, L. & Mitchell, A.P. Genes Dev 85, 533–548 (1991).	YSCPKN
113	PBS2	Encodes a member of the Ste7/MAP kinase kinase family. Overexpression confers resistance to the antibiotic polymixin B and improves the viability of cells containing dominant activating mutations of Ras.	Boguslawski, G. J. Gen. Microbiol. 138, 2425–2432 (1992).	YSCPBS2
114	CKA1	CKA1 and CKA2 both encode catalytic subunits of casein kinase II. Neither gene is essential, but deletion of both is lethal.	Padmanabha, R., Chen, W.J., Hanna, D.E. & Glover, C.V. Mol Cell Biol 10, 4089–4099 (1990).	YSCCSKAS

Entry	Gene	Description	Reference	Genbank Name
115	CKA2	See CKA1.		YSCCKA2
116	YCK1	YCK1 and YCK2 both encode casein kinase I. Neither gene is essential, but deletion of both is lethal.	Robinson, L.C., et al. Proc Natl Acad Sci U S A **89**, 28–32 (1992).	YSCYCK1
117	YCK2	See YCK1.		YSCYCK2
118	HRR25	Encodes a homolog of casein kinase I. Δhrr25 cells are viable but are defective in their ability to repair DNA damage.	DeMaggio, A.J., Lindberg, R.A., Hunter, T. & Hoekstra, M.F. Proc Natl Acad Sci U S A **89**, 7008–7012 (1992).	YSCHRR25A
119	DBF2	Encodes a protein kinase. Mutants delay DNA synthesis and appear to arrest in mitosis. Interacts genetically with SPO12 and SIT4.	Parkes, V. & Johnston, L.H. Nucleic Acids Res **20**, 5617–5623 (1992).	YSCDBF2A
120	DBF20	Encodes a protein kinase homologous to Dbf2. Neither gene is essential, but deletion of both is lethal.	Toyn, J.H., Araki, H., Sugino, A. & Johnston, L.H. Gene **104**, 63–70 (1991).	YSCDBFA

DNA Replication

Entry	Gene	Description	Reference	Genbank Name
121	POL1	Also called CDC17 (16).		
122	POL2	Encodes the catalytic subunit of DNA polymerase ε (also called polymerase III). Essential.	Araki, H., et al. Embo J **11**, 733–740 (1992).	YSCDNAPOL
123	POL3	Also called CDC2 (1).		
124	POL30	Encodes the yeast homolog of PCNA (proliferating cell nuclear antigen), an accessory factor for DNA polymerase δ required for leading strand synthesis. Essential.	Burgers, P.M. J Biol Chem **266**, 22698–22706 (1991).	YSCPOL30
125	POB1	Also called CHL15 and CTF4. Encodes an accessory protein that binds to DNA polymerase α. Δpob1cells are viable but have increased frequencies of chromosome loss.	Miles, J. & Formosa, T. Mol Cell Biol **12**, 5724–5735 (1992).	YSCPOB1A
126	DPB2	Encodes a subunit of DNA polymerase ε. Essential.	Araki, H., Hamatake, R.K., Johnston, L.H. & Sugino, A. Proc Natl Acad Sci U S A **88**, 4601–4605 (1991).	YSCDPB2
127	DPB3	Encodes a subunit of DNA polymerase ε. Nonessential.	Araki, H., et al. Nucleic Acids Res **19**, 4867–4872 (1991).	YSCDPB3G
128	PRI1	PRI1 and PRI2 encode the two subunits of DNA primase, the enzyme that synthesizes the RNA primers required for DNA synthesis. Both genes are essential.	Santocanale, C., et al. Gene **113**, 199–205 (1992).	SCPRI1

Entry	Gene	Description	Reference	Genbank Name
129	*PRI2*	See *PRI1*.		YSCPRI2
130	*RNR1*	*RNR1* and *RNR3* both encode regulatory subunits for ribonucleotide reductase, the enzyme that converts ribonucleotides to deoxyribonucleotides. *RNR2* encodes the catalytic subunit. *RNR1* and *RNR2* are essential.	Elledge, S.J., Zhou, Z. & Allen, J.B. *Trends Biochem Sci* **17**, 119–123 (1992).	
131	*RNR2*	See *RNR1*.		YSCRNR2A
132	*RNR3*	Also called *DIN1*. See *RNR1*. Not essential. Strongly induced by DNA damage.	Zhou, Z. & Elledge, S.J. *Genetics* **131**, 851–866 (1992).	YSCDINI
133	*MCM2*	Missense mutations in *MCM* genes differentially affect replication from different replication origins. *MCM2* and *MCM3* are both essential and both encode proteins related to Cdc46. *MCM2* interacts genetically with *MCM3*.	Yan, H., Gibson, S. & Tye, B.K. *Genes Dev* **5**, 944–957 (1991).	YSCMCM2
134	*MCM3*	See *MCM2*.		YSCMCM3G
135	*MCM5*	Also called *CDC46* (39).		
136	*RFA1*	Also called *RPA1*. *RFA1, RFA2,* and *RFA3* encode the 70-kd, 34-kd, and 13-kd subunits of replication protein A (RP-A), which binds to single-stranded DNA. All three genes are essential. Rfa2 is phosphorylated in a cell cycle-dependent manner by Cdc28–cyclin complexes.	Brill, S.J. & Stillman, B. *Genes Dev* **15**, 1589–1600 (1991).	YSCRFA1
137	*RFA2*	See *RFA1*.		YSCRFA2
138	*RFA3*	See *RFA1*.		YSCRFA3
139	*ORC1*	*ORC1* through *ORC6* encode the six subunits of the origin recognition complex (ORC).	Bell, S.P. & Stillman, B. *Nature* **357**, 128–134 (1992).	
140	*ORC2*	See *ORC1*. Encodes the 72-kd subunit of ORC. Essential.		
141	*ORC3*	See *ORC1*. Encodes the 62-kd subunit of ORC.		
142	*ORC4*	See *ORC1*. Encodes the 56-kd subunit of ORC.		
143	*ORC5*	See *ORC1*. Encodes the 53-kd subunit of ORC.		
144	*ORC6*	See *ORC1*. Also called *AAP1*. Encodes the 50-kd subunit of ORC. Essential.		
145	*ABF1*	Encodes a transcriptional regulator that binds to *ARS1,* a replication origin. The Abf1 binding site in *ARS1* can be replaced by sites for other transcriptional regulators. Essential.	Marahrens, Y. & Stillman, B. *Science* **255**, 817–823 (1992).	YSCABF1A

Entry	Gene	Description	Reference	Genbank Name
146	DBF4	Sequenced but not homologous to known genes. Required for the initiation of DNA synthesis. Interacts genetically with CDC7.	Kitada, K., Johnston, L.H., Sugino, T. & Sugino, A. *Genetics* **131**, 21–29 (1992).	YSCDBF4
147	HTA1	HTA1 and HTA2 both encode histone H2A. Neither gene is essential, but deletion of both is lethal.	Osley, M.A. *Annu Rev Biochem* **60**, 827–861 (1991).	YSCH2A1
148	HTA2	See HTA1.		YSCH2A2
149	HTB1	HTB1 and HTB2 both encode histone H2B. Neither gene is essential, but deletion of both is lethal.	Osley, M.A. *Annu Rev Biochem* **60**, 827–861 (1991).	YSCH2B1
150	HTB2	See HTB1.		YSCH2B2
151	HHT1	HHT1 and HHT2 both encode histone H3. Neither gene is essential, but deletion of both is lethal.	Mann, R.K. & Grunstein, M. *Embo J* **11**, 3297–3306 (1992).	YSCH34CI
152	HHT2	See HHT1.		YSCH34CII
153	HHF1	HHF1 and HHF2 both encode histone H4. Neither gene is essential, but deletion of both is lethal.	Johnson, L.M., Fisher, A.G. & Grunstein, M. *Embo J* **11**, 2201–2209 (1992).	YSCH34CI
154	HHF2	See HHF1.		YSCH34CII
155	TOP1	Encodes DNA topoisomerase I. Not essential. Cells require either topoisomerase I or topoisomerase II activity during DNA replication.	Choder, M. *Genes Dev* **5**, 2315–2326 (1991).	YSCTOPI
156	TOP2	Encodes DNA topoisomerase II. Essential for sister chromatid segregation. See TOP1.	Thomas, W., Spell, R.M., Ming, M.E. & Holm, C. *Genetics* **128**, 703–716 (1991).	YSCTOP2
157	TOP3	Encodes a type I topoisomerase with properties similar to prokaryotic type I topoisomerases. Not essential, but Δtop3 cells proliferate slowly and have increased recombination between repeated sequences.	Kim, R.A. & Wang, J.C. *J Biol Chem* **267**, 17178–17185 (1992).	YSCTOP3

Spindle components

Entry	Gene	Description	Reference	Genbank Name
158	TUB1	TUB1 encodes the major form of α tubulin, α1, and TUB3 encodes the minor form, α2. TUB1 is essential but TUB3 is not.	Stearns, T., Hoyt, M.A. & Botstein, D. *Genetics* **124**, 251–262 (1990).	YSCTUB1
159	TUB2	Encodes β tubulin. Essential.	Sullivan, D.S. & Huffaker, T.C. *J Cell Biol* **119**, 379–388 (1992).	YSCTUBB
160	TUB3	See TUB1.		YSCTUB3

Entry	Gene	Description	Reference	Genbank Name
161	CIN1	Cloned but not homologous to known genes. Deletion of CIN1, CIN2, or CIN4 makes microtubules much more sensitive to antimicrotubule drugs. Analyzing the interaction of these mutants with each other and tubulin mutants suggests that Cin1, Cin2, and Cin4 act together in a pathway that regulates the stability of microtubules.	Stearns, T., Hoyt, M.A. & Botstein, D. *Genetics* **124,** 251–262 (1990).	
162	CIN2	Cloned but not homologous to known genes. See *CIN1.*		
163	CIN4	Encodes a small G protein. See *CIN1.*		
164	KAR1	Encodes a component of the spindle pole body that appears to regulate microtubule function. Essential. *kar1⁻* mutants cannot perform nuclear fusion during mating.	Vallen, E.A., Hiller, M.A., Scherson, T.Y. & Rose, M.D. *J Cell Biol* **117,** 1277–1287 (1992).	YSCKARI
165	KAR3	Encodes a member of the kinesin family. Δ*kar3* cells are delayed in mitosis and cannot perform nuclear fusion during mating.	Roof, D.M., Meluh, P.B. & Rose, M.D. *J Cell Biol* **118,** 95–108 (1992).	YSCKAR3AA
166	CIN8	Also called *KSL2. CIN8* and *KIP1* both encode heavy chains of members of the kinesin family. Neither gene is essential, but deletion of both is lethal and prevents separation of the spindle pole bodies. Genetic interactions with *KAR3* (165) suggest that Kar3 generates spindle forces in the opposite direction from those due to Cin8 and Kip1.	Saunders, W.S. & Hoyt, M.A. *Cell* **70,** 451–458 (1992).	YSCCIN8A
167	KIP1	Also called *CIN9.* See *CIN8.*	Roof, D.M., Meluh, P.B. & Rose, M.D. *J Cell Biol* **118,** 95–108 (1992).	SCKIPIG
168	KIP2	Encodes the heavy chain of a member of the kinesin family. Not essential.	Roof, D.M., Meluh, P.B. & Rose, M.D. *J Cell Biol* **118,** 95–108 (1992).	YSCKIP2IIL
169	MPS1	Encodes a protein kinase. *ts* mutants block spindle pole body duplication but do not arrest in mitosis, suggesting that they are defective in the feedback control over the exit from mitosis.	Winey, M., Goetsch, L., Baum, P. & Byers, B. *J Cell Biol* **114,** 745–754 (1991).	
170	MPS2	Cloned but not sequenced. *ts* mutants cause formation of a defective spindle pole body that fails to nucleate microtubules. *ts* mutants delay in mitosis.	Winey, M., Goetsch, L., Baum, P. & Byers, B. *J Cell Biol* **114,** 745–754 (1991).	

Entry	Gene	Description	Reference	Genbank Name
171	*ESP1*	Encodes a large protein that has regions of homology to the fission yeast *cut1* and Aspergillus *bimB* genes. *ts* mutants form abnormal spindles and missegregate their spindle pole bodies.	Baum, P., Yip, C., Goetsch, L. & Byers, B. *Mol Cell Biol* **8**, 5386–5397 (1988).	YSCESPIA
172	*NDC1*	Cloned. *ts* mutants cause formation of a defective spindle pole body and segregate all their chromosomes to one pole at mitosis.	Vallen, E.A., *et al. Cell* **69**, 505–515 (1992).	SCNDC1A
173	*CIK1*	Encodes a protein that is localized at the spindle pole. Not essential for viability but required for nuclear fusion during mating.	Page, B.D. & Snyder, M. *Genes Dev* **6**, 1414–1429 (1992).	YSCCIKI
174	*CBF1*	Encodes a protein that binds to element I of the centromere sequence as well as the promoters of genes involved in methionine biosynthesis. Δ*cbf1* cells are viable but show increased frequencies of chromosome loss and require exogenous methionine.	Cai, M. & Davis, R.W. *Cell* **61**, 437–446 (1990).	YSCCFB1A
175	*CTF13*	Encodes a 56-kd subunit of the protein complex that binds to element III of the yeast centromere DNA. Essential.	Spencer, F., Gerring, S.L., Connelly, C. & Hieter, P. *Genetics* **124**, 237–249 (1990).	
176	*CTF14*	Also called *NDC10* and *CBF2*. Encodes a 110-kd subunit of the protein complex that binds to element III of the yeast centromere DNA. Essential. *ts* mutants arrest in mitosis.	Spencer, F., Gerring, S.L., Connelly, C. & Hieter, P. *Genetics* **124**, 237–249 (1990).	
177	*CHL1*	Encodes a protein with homology to known helicases. Δ*chl1* cells are viable but show increased frequencies of chromosome loss and interact with mutations in *mad1* and *mad2*.	Gerring, S.L., Spencer, F. & Hieter, P. *Embo J* **9**, 4347–4358 (1990).	YSCCHL1

Budding and actin cytoskeleton

Entry	Gene	Description	Reference	Genbank Name
178	*BUD1*	Also called *RSR1*. Encodes a small G protein required for correct bud site selection.	Ruggieri, R., *et al. Mol Cell Biol* **12**, 758–766 (1992).	YSCRSR1
179	*BUD2*	Encodes a GTPase activating protein that acts on Bud1. Required for correct bud site selection.	Chant, J. & Herskowitz, I. *Cell* **65**, 1203–1212 (1991).	
180	*BUD3*	Encodes a protein that is localized to the bud site in diploid cells. Required for correct bud site selection.	Chant, J. & Herskowitz, I. *Cell* **65**, 1203–1212 (1991).	

Entry	Gene	Description	Reference	Genbank Name
181	*BUD4*	Not cloned. Required for correct bud site selection.	Chant, J. & Herskowitz, I. *Cell* **65**, 1203–1212 (1991).	
182	*BUD5*	Encodes a protein with homology to guanine nucleotide exchange proteins. Required for correct bud site selection.	Chant, J., Corrado, K., Pringle, J.R. & Herskowitz, I. *Cell* **65**, 1213–1224 (1991).	YSCBUD
183	*BEM1*	Encodes a protein that contains SH3 domains (implicated in interactions with the actin cytoskeleton). Not essential, but deletions are partially defective in bud emergence. Interacts genetically with *BUD5* and *MSB1*.	Chenevert, J., *et al. Nature* **356**, 77–79 (1992).	YSCBEM1G
184	*BEM2*	Not cloned. Probable role in bud emergence. Interacts genetically with *CDC55* and *MSB1*.	Bender, A. & Pringle, J.R. *Mol Cell Biol* **11**, 1295–1305 (1991).	
185	*MSB1*	Cloned but not homologous to known genes. Not essential. Interacts genetically with *CDC24, CDC42,* and *BEM1*.	Bender, A. & Pringle, J.R. *Mol Cell Biol* **11**, 1295–1305 (1991).	YSCMSB1R1
186	*MSB2*	Cloned but not homologous to known genes. Not essential. Suppresses *cdc24*[ts].	Bender, A. & Pringle, J.R. *Yeast* **8**, 315–323 (1992).	YSCMSB2A
187	*SPA2*	Cloned but not homologous to known genes. Δ*spa2* x Δ*spa2* crosses have reduced mating efficiency. Cells are born with Spa2 localized at the point on their surface at which a bud will emerge.	Costigan, C., Gehrung, S. & Snyder, M. *Mol Cell Biol* **12**, 1162–1178 (1992).	YSCSPA2G
188	*SLK1*	Encodes a protein kinase whose deletion is lethal in Δ*spa2* cells. Δ*slk1* cells are *ts* for proliferation.	Costigan, C., Gehrung, S. & Snyder, M. *Mol Cell Biol* **12**, 1162–1178 (1992).	YSCSLK1A
189	*ACT1*	Encodes actin. Essential. *ts* mutants show defects in bud emergence and bud growth.	Johannes, F.J. & Gallwitz, D. *Embo J* **10**, 3951–3958 (1991).	YSPACT1
190	*ACT2*	Encodes a protein related to actin. Essential.	Schwob, E. & Martin, R.P. *Nature* **355**, 179–182 (1992).	YSCACT2G
191	*ABP1*	Encodes an actin-binding protein that contains an SH3 domain. Not essential.	Read, E.B., Okamura, H.H. & Drubin, D.G. *Mol Biol Cell* **3**, 429–444 (1992).	SCABP1
192	*CAP1*	*CAP1* encodes the α subunit and *CAP2* encodes the β subunit of an actin-binding protein. Neither gene is essential, but deletion of either one or both causes abnormal actin distribution and changes in cell morphology.	Amatruda, J.F., Gattermeir, D.J., Karpova, T.S. & Cooper, J.A. *J Cell Biol* **119**, 1151–1162 (1992).	YSCCAP1G
193	*CAP2*	See *CAP1*.		YSCCAP2

Entry	Gene	Description	Reference	Genbank Name
194	SAC6	Encodes an actin-binding protein homologous to fimbrin, an actin-bundling protein. Not essential, but Δsac6 cells show abnormal actin distribution.	Adams, A.E., Botstein, D. & Drubin, D.G. *Nature* **354**, 404–408 (1991).	
195	MYO1	Encodes a type II myosin heavy chain. Not essential.	Rodriguez, J.R. & Paterson, B.M. *Cell Motil Cytoskeleton* **17**, 301–308 (1990).	YSCMYO1G
196	MYO2	Encodes a tailless myosin. Essential. *ts* mutants block bud emergence.	Johnston, G.C., Prendergast, J.A. & Singer, R.A. *J Cell Biol* **113**, 539–551 (1991).	YSCMYO2A
197	MYO4	Encodes a relative of *MYO2*. Not essential.		YSCMYO4P
198	SMY2	Encodes a protein with homology to the motor domain of kinesins. Overexpression suppresses *myo2ts*.	Lillie, S.H. & Brown, S.S. *Nature* **356**, 358–361 (1992).	YSCSMY2P
199	TPM1	Encodes a homolog of tropomyosin. Not essential, but deletion disrupts actin organization and reduces mating efficiency. Interacts genetically with *MYO2*.	Liu, H. & Bretscher, A. *J Cell Biol* **118**, 285–299 (1992).	YSCTRO1

Checkpoint/feedback controls

Entry	Gene	Description	Reference	Genbank Name
200	RAD9	Cloned but not homologous to known genes. *rad9⁻*, *rad17⁻*, *rad24⁻*, and *mec3⁻* mutants inactivate the feedback control that prevents nuclear division in cells that contain damaged DNA. Δ*rad9* cells are viable.	Weinert, T.A. *Radiat Res* **132**, 141–143 (1992).	YSCRAD9
201	RAD17	Not cloned. See *RAD9*.		
202	RAD24	Not cloned. See *RAD9*.		
203	MEC1	Not cloned. *mec1⁻* and *mec2⁻* mutants inactivate the feedback control that prevents nuclear division in cells that contain damaged or unreplicated DNA.	Weinert, T.A. *Radiat Res* **132**, 141–143 (1992).	
204	MEC2	Not cloned. See *MEC1*.		
205	MEC3	Not cloned. See *RAD9* (200).		
206	BUB1	Encodes a putative protein kinase. *bub1⁻*, *bub2⁻* and *bub3⁻* mutations inactivate the feedback control that keeps cells with improperly assembled spindles from leaving mitosis. *BUB1* is not essential.	Hoyt, M.A., Totis, L. & Roberts, B.T. *Cell* **66**, 507–517 (1991).	
207	BUB2	Cloned but not homologous to known genes. Not essential. See *BUB1*.		YSCBUB2Q
208	BUB3	Cloned but not homologous to known genes. Not essential. See *BUB1*.		YSCBUB3Q

Entry	Gene	Description	Reference	Genbank Name
209	MAD1	Not cloned. mad1⁻, mad2⁻, and mad3⁻ mutations inactivate the feedback control that keeps cells with improperly assembled spindles from leaving mitosis.	Li, R. & Murray, A.W. *Cell* **66**, 519–531 (1991).	
210	MAD2	Encodes the α subunit of a prenyl transferase. Essential. See MAD1.		YSCMAD2
211	MAD3	Not cloned. See MAD1.		

Mating

Entry	Gene	Description	Reference	Genbank Name
212	MATα1	Part of the MATα mating-type locus. Encodes a transcriptional activator that cooperates with Mcm1 to induce α-specific genes (genes expressed only in α cells).	Sengupta, P. & Cochran, B.H. *Genes Dev* **5**, 1924–1934 (1991).	YSCMATLOC
213	MATα2	Part of the MATα mating-type locus. Encodes a DNA-binding protein that interacts with Mcm1 to repress **a**-specific genes (genes expressed only in **a** cells). In diploid cells Matα2 interacts with Mata1 to repress haploid-specific genes (genes expressed only in haploid cells).	Smith, D.L. & Johnson, A.D. *Cell* **68**, 133–142 (1992).	YSCMATLOC
214	MATa1	Part of the MATa mating type locus. See MATα2.	Nakazawa, N., Harashima, S. & Oshima, Y. *Mol Cell Biol* **11**, 5693–5700 (1991).	YSCMATA
215	MCM1	Encodes a DNA-binding protein that forms part of various different transcription complexes. Essential. Missense mutants are defective in initiating replication at some origins.	Smith, D.L. & Johnson, A.D. *Cell* **68**, 133–142 (1992).	YSCMCM1
216	HO	Encodes the restriction endonuclease that initiates mating-type switching. Expressed coordinately with CLN1 and CLN2.	Dohrmann, P.R., *et al.* *Genes Dev* **6**, 93–104 (1992).	YSCHORR
217	SWI5	Encodes a transcription factor that acts as a licensing factor for HO expression. Swi5 is synthesized during G2, but Cdc28-catalyzed phosphorylation keeps it from entering the nucleus until anaphase.	Moll, T., *et al. Cell* **66**, 743–758 (1991).	YSCSW15
218	STE2	STE2 encodes the α factor receptor and STE3 encodes the **a** factor receptor. Mating factor binding stimulates activation of the Ste4-containing trimeric G protein.	Konopka, J.B. & Jenness, D.D. *Cell Regul* **2**, 439–452 (1991).	YSCSTE2G

Entry	Gene	Description	Reference	Genbank Name
219	STE3	See STE2.	Hagen, D.C., McCaffrey, G. & Sprague, G.J. *Proc Natl Acad Sci U S A* **83**, 1418–1422 (1986).	YSCSTE3G
220	STE4	Encodes the β subunit of the trimeric G protein (Scg1/Ste4/Ste18) that interacts with the mating factor receptors. The β–γ complex activates the signaling pathway when it is not bound to the a subunit.	Whiteway, M., Hougan, L. & Thomas, D.Y. *Mol Cell Biol* **10**, 217–222 (1990).	YSCSTE4
221	STE5	Cloned but not homologous to known genes. Required for the activation of Ste11.	Stevenson, B.J., Rhodes, N., Errede, B. & Sprague, G.J. *Genes Dev* **6**, 1293–1304 (1992).	YSCSTE5
222	STE6	Encodes a protein that is required for the export of a factor from the cell. Ste6 is homologous to the mammalian multidrug transporter.	Berkower, C. & Michaelis, S. *Embo J* **10**, 3777–3785 (1991).	YSCSTE6A
223	STE7	Encodes a protein kinase homologous to MAP kinase kinase. Phosphorylated and activated by Ste11.	Cairns, B.R., Ramer, S.W. & Kornberg, R.D. *Genes Dev* **6**, 1305–18 (1992).	YSCSTE7
224	STE11	Encodes a protein kinase with some general similarities to Raf. Probably activated by phosphorylation. Ste11 phosphorylates and activates Ste7.	Stevenson, B.J., Rhodes, N., Errede, B. & Sprague, G.J. *Genes Dev* **6**, 1293–1304 (1992).	YSCSTE11
225	STE12	Encodes a transcription factor required for expression of genes in the signaling and G1 arrest pathways. Probably phosphorylated by either Fus3 or Kss1.	Yuan, Y.L. & Fields, S. *Mol Cell Biol* **11**, 5910–5918 (1991).	YSCSTE12A
226	STE13	Encodes a protease (dipeptidylaminopeptidase) required for production of mature α factor.	Julius, D., *et al. Cell* **32**, 839–852 (1983).	
227	STE14	Encodes the enzyme that methylates the carboxyl terminal of **a** factor after it has been prenylated and proteolytically trimmed.	Hrycyna, C.A., Sapperstein, S.K., Clarke, S. & Michaelis, S. *Embo J* **10**, 1699–1709 (1991).	
228	STE16	Also called *RAM2* (86).		
229	STE18	Encodes the γ subunit of the trimeric G protein that interacts with the mating factor receptors. See *STE4*.	Whiteway, M., *et al. Cell* **56**, 467–477 (1989).	YSCSTE18
230	STE20	Encodes a protein kinase with homology to protein kinase C. Ste20 is probably activated by interacting with Ste4.	Leberer, E., *et al. EMBO J.* **11**, 4815–4824 (1992).	YSCSERKIN

Entry	Gene	Description	Reference	Genbank Name
231	SCG1	Also called GPA1. Encodes the α subunit of the trimeric G protein that is coupled to the mating factor receptors.	Hirsch, J.P., Dietzel, C. & Kurjan, J. Genes Dev 5, 467–474 (1991).	YSTSCG1A
232	NMT1	Also called CDC72 (53).		
233	FUS1	Encodes a membrane bound and glycosylated protein that is required for cell fusion during mating. Strongly induced by mating factors.	Trueheart, J. & Fink, G.R. Proc Natl Acad Sci U S A 86, 9916–9920 (1989).	YSCFUS1G
234	FUS2	Not cloned. Required for cell fusion during mating.	Trueheart, J., Boeke, J.D. & Fink, G.R. Mol Cell Biol 7, 2316–2328 (1987).	
235	FUS3	Also called DAC2. Encodes a member of the MAP kinase family. Fus3 phosphorylates and activates Far1 and is also implicated in the modification of Cln3 induced by mating factor treatment. Either Kss1 or Fus3 are required to activate Ste12.	Gartner, A., Nasmyth, K. & Ammerer, G. Genes Dev 6, 1280–1292 (1992).	YSCFUS3
236	KSS1	Encodes a member of the MAP kinase family. See FUS3.		YSCPKS
237	FAR1	Cloned but not homologous to known genes. Δfar1 cells fail to arrest when treated with mating factors, but can still induce mating genes. Far1 binds to Cln1 and Cln2 and is required to mediate their mating-factor-induced destruction.	Chang, F. & Herskowitz, I. Mol Biol Cell 3, 445–450 (1992).	YSCFAR1
238	DAF2	Not cloned. A dominant mutation weakly induces the transcription of mating genes in the absence of mating factors and prevents the cell cycle arrest caused by overexpression of STE4.	Cross, F.R. Genetics 126, 301–308 (1990).	
239	SST2	Cloned but not homologous to known genes. Mutants are defective in recovery from exposure to either a or α factor.	Dietzel, C. & Kurjan, J. Mol Cell Biol 7, 4169–4177 (1987).	YSCSST2
240	SGV1	Encodes a protein kinase with homology to Cdc2/28. Mutants slow adaptation to mating factors and cause cs and ts proliferation.	Irie, K., Nomoto, S., Miyajima, I. & Matsumoto, K. Cell 65, 785–795 (1991).	YSCSGV1

Entry	Gene	Description	Reference	Genbank Name
241	SRM1	Also called *PRP20*. Encodes a guanine nucleotide exchange protein homologous to the mammalian RCC1 (repressor of chromosome condensation) protein. Mutants allow cells without mating factor receptor to mate, and at 37°C prevent RNA splicing. No known defect in feedback control.	Clark, K.L., *et al. Cell Regul* **2**, 781–792 (1991).	M27013
242	MFα1	*MFα1* and *MFα2* both encode precursors of α factor. Deletion of both genes is required to abolish the production of α factor.	Caplan, S. & Kurjan, J. *Genetics* **127**, 299–307 (1991).	YSCMFA
243	MFα2	See *MFα1*.		YSCMFA2G
244	KEX1	Encodes a carboxypeptidase involved in the production of α factor.	Cooper, A. & Bussey, H. *Mol Cell Biol* **9**, 2706–2714 (1989).	YSCKEX1
245	KEX2	Encodes an endopeptidase involved in the production of α factor.	Brenner, C. & Fuller, R.S. *Proc Natl Acad Sci U S A* **89**, 922–926 (1992).	YSCKEX2A
246	BAR1	Also called *SST1*. Encodes an extracellular protease degrades α factor. Only expressed in **a** cells. *bar1⁻* **a** cells recover slowly from exposure to α factor.	MacKay, V.L., *et al. Proc Natl Acad Sci U S A* **85**, 55–59 (1988).	YSCBAR1A
247	MFA1	*MFA1* and *MFA2* both encode precursors of **a** factor. Deletion of both genes is required to abolish the production of **a** factor.		
248	MFA2	See *MFA1*.		
249	SSL1	Cloned but not sequenced. Mutants are hypersensitive to **a** factor, and the gene is expressed only in α cells. (Two other unrelated genes are also named *SSL1*.)	Steden, M., Betz, R. & Duntze, W. *Mol Gen Genet* **219**, 439–444 (1989).	
250	AGα1	Encodes the agglutinin molecule expressed on α cells that allows them to bind to **a** cells during mating. Not essential for mating.	Lipke, P.N., Wojciechowicz, D. & Kurjan, J. *Mol Cell Biol* **9**, 3155–3165 (1989).	YSCAGA1A
251	AGA1	*AGA1* and *AGA2* encode the two subunits of the agglutinin molecule expressed on **a** cells that allows them to bind to α cells during mating. Aga1 is the core subunit attached to the surface of cells and Aga2 is the binding subunit that interacts with α cells. Neither gene is essential for mating.	Lipke, P.N. & Kurjan, J. *Microbiol Rev* **56**, 180–194 (1992).	YSCAAGLCS
252	AGA2	See *AGA1*.		SCAGA2MR

Entry	Gene	Description	Reference	Genbank Name

Meiosis

253	SPO7	Cloned but not homologous to known genes. Essential for premeiotic DNA synthesis but not for DNA synthesis in mitotic cell cycles.	Whyte, W., et al. Gene **95**, 65–72 (1990).	YSCSPO7A
254	SPO11	Cloned but not homologous to known genes. Essential for synaptonemal complex formation and the initiation of meiotic recombination.	Cao, L., Alani, E. & Kleckner, N. Cell **61**, 1089–1101 (1990).	YSCSPO11
255	SPO12	Cloned but not homologous to known genes. Mutants produce only two spores. Most chromosomes segregate as they would in meiosis I, but some segregate as they would in meiosis II.	Malavasic, M.J. & Elder, R.T. Mol Cell Biol **10**, 2809–2819 (1990).	YSCSPOA
256	SPO13	Cloned but not homologous to known genes. Δspo13 cells produce only two spores as a result of a failure to perform meiosis I. Ectopic expression arrests the mitotic cell cycle in G2/mitosis.	Buckingham, L.E., et al. Proc Natl Acad Sci U S A **87**, 9406–9410 (1990).	YSCSPO13
257	SPO14	Not cloned. Δspo14 cells proceed normally through meiosis I but have a defect in meiosis II and are unable to form spores.	Honigberg, S.M., Conicella, C. & Espositio, R.E. Genetics **130**, 703–716 (1992).	
258	SPO15	Also called VPS1. Encodes a homolog of dynamin, a microtubule-bundling protein. SPO15 is essential for spindle pole separation in meiosis and correct sorting of proteins to the vacuole in the mitotic and meiotic cell cycles.	Yeh, E., et al. Nature **349**, 713–715 (1991).	YSCSPO15G
259	SPO16	Sequenced but lacks detectable homology to known genes. Δspo16 cells fail to sporulate.	Malavasic, M.J. & Elder, R.T. Mol Cell Biol **10**, 2809–2819 (1990).	YSCSPOA
260	RAD50	Encodes a putative nucleotide-binding protein with coiled coil regions. Δrad50 cells fail to initiate meiotic recombination or synapse homologous chromosomes. One missense mutant (rad50^S) forms double-stranded breaks during recombination but fails to convert them into recombination events. Required for repair of double-strand breaks in the mitotic cell cycle.	Alani, E., Padmore, R. & Kleckner, N. Cell **61**, 419–436 (1990).	YSCRAD50

Entry	Gene	Description	Reference	Genbank Name
261	*RAD51*	Encodes a DNA binding protein with homology to the bacterial RecA protein. Required for meiotic recombination and repair of DNA damage in the mitotic cell cycle.	Shinohara, A., Ogawa, H. & Ogawa, T. *Cell* **69**, 457–470 (1992).	YSCRAD51
262	*RAD52*	Cloned but not homologous to known genes. Required for the repair of double-stranded breaks in the mitotic cell cycle and the processing of recombination intermediates in meiotic cells. Not required for synaptonemal complex formation.	Dornfeld, K.J. & Livingston, D.M. *Mol Cell Biol* **11**, 2013–2017 (1991).	YSCRAD52
263	*RAD54*	Encodes a protein with regions of homology to other repair functions and some transcriptional activators. Required for repair of double-stranded breaks in the mitotic cell cycle and recombination in meiotic ones.	Emery, H.S., Schild, D., Kellogg, D.E. & Mortimer, R.K. *Gene* **104**, 103–106 (1991).	YSCRAD54A
264	*RAD57*	Encodes a protein with homology to Rad51. Required for repair of double-stranded breaks in the mitotic cell cycle and recombination in meiotic ones.	Kans, J.A. & Mortimer, R.K. *Gene* **105**, 139–140 (1991).	YSCRAD57A
265	*DMC1*	Encodes a protein with homology to Rad51. Mutants fail to form full synaptonemal complexes, accumulate unprocessed recombination intermediates, and arrest in interphase of the meiotic cell cycle.	Bishop, D.K., Park, D., Xu, L. & Kleckner, N. *Cell* **69**, 439–456 (1992).	YSCDMC1A
266	*SEP1*	Also called *KEM1* and *XRN1*. Encodes an exonuclease and DNA strand-pairing activity. Essential for meiosis.	Tishkoff, D.X., Johnson, A.W. & Kolodner, R.D. *Mol Cell Biol* **11**, 2593–2608 (1991).	YSCSEP1
267	*XRS2*	Not cloned. Required for the initiation of meiotic recombination and repair of double-stranded breaks in the mitotic cell cycle.	Ivanov, E.L., Korolev, V.G. & Fabre, F. *Genetics* **132**, 651–664 (1992).	
268	*MER1*	Encodes a protein that is required for the correct splicing of Mer2 mRNA. Δ*mer1* cells form axial elements of synaptonemal complex, but chromosomes do not synapse. Recombination occurs at 10–20% of the wild-type frequency. Not required in the mitotic cell cycle.	Engebrecht, J.A., Voelkel, M.K. & Roeder, G.S. *Cell* **66**, 1257–1268 (1991).	YSCMER1

Entry	Gene	Description	Reference	Genbank Name
269	MER2	Cloned but not homologous to known proteins. Δmer2 cells form short stretches of the axial elements of synaptonemal complex, but chromosomes do not synapse or initiate recombination. Not required in the mitotic cell cycle.	Engebrecht, J.A., Voelkel, M.K. & Roeder, G.S. *Cell* **66**, 1257–1268 (1991).	YSCMER2A
270	MEI4	Cloned but not homologous to known genes. Δmei4 cells have the same phenotype as Δmer2 cells. Not required in the mitotic cell cycle.	Menees, T.M., Ross, M.P. & Roeder, G.S. *Mol Cell Biol* **12**, 1340–1351 (1992).	YSCMEI4B
271	REC102	Cloned but not homologous to known genes. Δrec102 cells have the same phenotype as Δmer2 cells. Not required in the mitotic cell cycle.	Cool, M. & Malone, R.E. *Mol Cell Biol* **12**, 1248–1256 (1992).	YSCREC102
272	HOP1	Encodes a protein with homology to DNA-binding proteins. Hop1 is found along meiotic chromosomes and hop1⁻ mutants fail to form synaptonemal complexes.	Hollingsworth, N.M., Goetsch, L. & Byers, B. *Cell* **61**, 73–84 (1990).	YSCHOP1
273	ZIP1	Encodes a protein with homology to known elongated proteins that dimerize. zip1⁻ mutants undergo normal recombination, but meiotic chromosomes are synapsed only at the sites of recombination. Zip1 is found along the length of meiotic chromosomes.	Sym, M., Engebrecht, J. & Roeder, G.S. *Cell* **72**, 365–378 (1993).	YSCZIP1A
274	MEK1	Also called MRE4. Encodes a protein kinase. Δmek1 cells have short stretches of synaptonemal complex and reduced levels of recombination and spore viability. This phenotype is rescued by spo13 mutants.	Rockmill, B. & Roeder, G.S. *Genes Dev* **5**, 2392–2404 (1991).	YSCMSPKH
275	RED1	Cloned but not homologous to known proteins. red1⁻ mutants recombine at 10–20% of the wild-type frequency but fail to form normal synaptonemal complex or segregate their chromosomes correctly in meiosis I.	Rockmill, B. & Roeder, G.S. *Genetics* **126**, 563–574 (1990).	YSCRED1
276	RED2	Not cloned. Mutants form normal synaptonemal complex and have only a twofold reduction in recombination, but they fail to form viable spores.	G. S. Roeder, personal communication	
277	DIS1	Not cloned. Dominant mutants cause premature sister centromere separation in meiosis I and chromosome nondisjunction in mitosis.	Rockmill, B. & Fogel, S. *Genetics* **119**, 261–272 (1988).	

Entry	Gene	Description	Reference	Genbank Name
278	SID1	Not cloned. A dominant mutant shows premature sister centromere separation in meiosis I and nondisjunction in meiosis I and meiosis II.	Flatters, M. & Dawson, D. *Genetics,* **in press** (1993).	
279	PMS1	Encodes a member of a family of proteins that are involved in the repair of mismatches between strands of DNA duplexes. *pms1⁻* cells are deficient in mismatch repair during meiosis.	Kramer, W., Kramer, B., Williamson, M.S. & Fogel, S. *J Bacteriol* **171**, 5339–5346 (1989).	YSCPMS1A
280	RME1	Encodes a repressor of *IME1* expression, which is expressed in haploid cells but not in diploid cells. Deletion mutations allow α/α or **a**/**a** diploids to sporulate.	Covitz, P.A., Herskowitz, I. & Mitchell, A.P. *Genes Dev* **5**, 1982–1989 (1991).	YSCZFP
281	IME1	Encodes a transcriptional regulator that is produced in **a**/α diploids in response to starvation. Δ*ime1* cells fail to sporulate. Expressing *IME1* from a heterologous promoter induces **a**/**a** or α/α cells to sporulate.	Smith, H.E., et al. Mol *Cell Biol* **10**, 6103–6113 (1990).	YSCIME1
282	IME2	Also called *SME1*. Encodes a protein kinase. *IME2* expression is induced by *IME1*. Expressing *IME2* from a heterologous promoter allows Δ*ime1* cells to sporulate.	Mitchell, A.P., Driscoll, S.E. & Smith, H.E. *Mol Cell Biol* **10**, 2104–2110 (1990).	YSCSMEIG
283	IME4	Encodes a protein required for the transcription of *IME1*. *IME4* transcription is induced in starved diploid cells.	Shah, J.C. & Clancy, M.J. *Mol Cell Biol* **12**, 1078–1086 (1992).	

Fission Yeast

CDC Genes

Entry	Gene	Description	Reference	Genbank Name
284	cdc1	Cloned but not homologous to known genes. *ts* mutants cannot enter mitosis.	Fantes, P.A., Warbrick, E., Hughes, D.A. & MacNeill, S.A. *Cold Spring Harb Symp Quant Biol* **56**, 605–611 (1991).	
285	cdc2	Encodes a protein kinase subunit that associates with cyclins to form active protein kinase complexes that induce passage through Start and mitosis.	Broek, D., Bartlett, R., Crawford, K. & Nurse, P. *Nature* **349**, 388–393 (1991).	YSPCDC2
286	cdc3	Encodes profilin. *ts* mutants fail to septate.	K. Gould, personal communication	

Entry	Gene	Description	Reference	Genbank Name
287	*cdc4*	Not cloned. *ts* mutants fail to septate.	Nurse, P., Thuriaux, P. & Nasmyth, K. *Mol Gen Genet* **146**, 167–178 (1976).	
288	*cdc5*	Encodes a protein with homology to c-Myb, a transcriptional activator. *ts* mutants cannot enter mitosis.	K. Gould, personal communication	
289	*cdc6*	Encodes the catalytic subunit of DNA polymerase ε (also called DNA polymerase III). *ts* mutants fail to complete DNA synthesis.	M. Yamamoto, personal communication	
290	*cdc7*	Encodes a protein kinase. *cdc7^ts^, cdc 11^ts^, cdc12^ts^, cdc14^ts^,* and *cdc15^ts^* mutants all fail to septate at 35°C. *cdc7, cdc11, cdc14,* and *cdc16* interact with each other genetically.	Marks, J., Fankhauser, C. & Simanis, V. *J Cell Sci* **101**, 801–808 (1992).	
291	cdc8	Encodes a form of tropomyosin. ts mutants fail to septate.	Balasubramanian, M.K., Helfman, D.M. & Hemmingsen, S.M. Nature 360, 84–7 (1992).	YSPTROP
292	*cdc10*	Encodes the fission yeast homolog of budding yeast Swi6. *ts* mutants fail to enter S phase. Cdc10 and Sct1 form a transcriptional activator that induces expression of genes transcribed during S phase.	Caligiuri, M. & Beach, D. *Cell* **72**, 607–619 (1993).	YSPCDC10
293	*cdc11*	Cloned but not homologous to known genes. See *cdc7*.		
294	*cdc12*	Cloned but not homologous to known genes. See *cdc7*.		
295	*cdc13*	Encodes cyclin B. Δ*cdc13* cells cannot enter mitosis, *ts* mutant shows chromosome condensation but cannot form a spindle.	Alfa, C.E., Ducommun, B., Beach, D. & Hyams, J.S. *Nature* **347**, 680–682 (1990).	YSPCDC13
296	*cdc14*	Cloned but not homologous to known genes. See *cdc7*.		
297	*cdc15*	Cloned but not homologous to known genes. See *cdc7*.		
298	*cdc16*	Encodes a homolog of budding yeast Bub2. *ts* mutants form multiple septa and anucleate cells and are defective in the feedback control over the exit from mitosis.	Marks, J., Fankhauser, C. & Simanis, V. *J Cell Sci* **101**, 801–808 (1992).	
299	*cdc17*	Encodes DNA ligase. *ts* mutants arrest in G2 with damaged DNA.	Barker, D.G., White, J.H. & Johnston, L.H. *Eur J Biochem* **162**, 659–67 (1987).	YSPCDC17
300	*cdc18*	Encodes a homolog of budding yeast Cdc6. *ts* mutants cannot enter S phase. Overexpression suppresses *cdc10^ts^*.	P. Nurse, personal communication	

Entry	Gene	Description	Reference	Genbank Name
301	cdc19	Not cloned. Required for DNA synthesis and/or mitosis.	Nasmyth, K. & Nurse, P. *Mol Gen Genet* **182,** 119–124 (1981).	
302	cdc20	Not cloned. Required for DNA synthesis.	Nasmyth, K. & Nurse, P. *Mol Gen Genet* **182,** 119–124 (1981).	
303	cdc21	Encodes a protein homologous to budding yeast Mcm2 and Mcm3 proteins. *ts* mutants arrest with a G2 DNA content.	Coxon, A., Maundrell, K. & Kearsey, S.E. *Nucleic Acids Res* **20,** 5571–5577 (1992).	YSPCDC21
304	cdc22	Encodes the regulatory subunit of ribonucleotide reductase. Required for DNA synthesis.	Lowndes, N.F., *et al. Nature* **355,** 449–453 (1992).	
305	cdc23	Not cloned. Required for DNA synthesis.	Nasmyth, K. & Nurse, P. *Mol Gen Genet* **182,** 119–124 (1981).	
306	cdc24	Not cloned. Required for DNA synthesis.	Nasmyth, K. & Nurse, P. *Mol Gen Genet* **182,** 119–124 (1981).	
307	cdc25	Encodes the major tyrosine phosphatase that removes the phosphate from tyrosine 15 of Cdc2. *ts* mutants cannot enter mitosis.	Dunphy, W.G. & Kumagai, A. *Cell* **67,** 189–196 (1991).	YSPCDC25
308	cdc27	Cloned but not homologous to known genes. *ts* mutants cannot enter mitosis.	Fantes, P.A., Warbrick, E., Hughes, D.A. & MacNeill, S.A. *Cold Spring Harb Symp Quant Biol* **56,** 605–611 (1991).	YSPCDC27B
309	cdc28	Not cloned. Required to enter mitosis.	Nasmyth, K. & Nurse, P. *Mol Gen Genet* **182,** 119–124 (1981).	

Regulators of CDC2

Entry	Gene	Description	Reference	Genbank Name
310	wee1	Encodes the major tyrosine kinase that phosphorylates tyrosine 15 of Cdc2. *ts* mutants enter mitosis at a smaller size than wild-type cells.	Parker, L.L. & Piwnica-Worms, H. *Science* **257,** 1955–1957 (1992).	YSPWEE1A
311	mik1	Encodes a homolog of Wee1 that also regulates the phosphorylation of Cdc2. Double Δ*mik1 wee1^ts* mutants undergo mitotic catastrophe at 35°C.	Lundgren, K., *et al. Cell* **64,** 1111–1122 (1991).	YSPMIK1
312	nim1	Also called *cdr1*. Encodes a protein kinase that phosphorylates and inactivates Wee1. Thought to be involved in control of Wee1 activity by nutrients.	P. Russell, personal communication	YSPCDR1G

Entry	Gene	Description	Reference	Genbank Name
313	cdr2	Not cloned. cdr2⁻ mutants fail to decrease cell size on poor medium and cdr2⁻ cdc25ᵗˢ double mutants proliferate very poorly.	Hudson, J.D., Feilotter, H. & Young, P.G. *Genetics* **126**, 309–315 (1990).	
314	stf1	Cloned but not homologous to known genes. Mutants can suppress Δcdc25. This suppression is abolished by overexpressing phosphatase 1 (Dis2).	Hudson, J.D., *et al. Cold Spring Harb Symp Quant Biol* **56**, 599–604 (1991).	
315	wis1	Encodes a protein kinase homolog. Deletion increases cell size and overexpression decreases cell size.	Warbrick, E. & Fantes, P.A. *Embo J* **10**, 4291–4299 (1991).	YSPWIS1
316	suc1	Encodes a small protein whose biochemical function is unknown. Isolated as a high copy suppressor of cdc2ᵗˢ mutations. Suc1 binds tightly to Cdc2 and other Cdk proteins.	Ducommun, B., Brambilla, P. & Draetta, G. *Mol Cell Biol* **11**, 6177–6184 (1991).	YSPSUC1
317	mcs1	Not cloned. mcs1⁻ and mcs2⁻ mutants suppress the mitotic catastrophe seen in cdc2–3wᴰ wee1ᵗˢ strains and are lethal when combined with cdc25ᵗˢ.	Molz, L., Booher, R., Young, P. & Beach, D. *Genetics* **122**, 773–782 (1989).	
318	mcs4	Not cloned. See mcs1.		
319	cig1	Encodes a B type cyclin. Δcig1 cells are viable but delayed in G1. Overexpression is lethal.	Bueno, A., Richardson, H., Reed, S.I. & Russell, P. *Cell* **66**, 149–159 (1991).	YSPBCYCLIN
320	puc1	Encodes a G1 cyclin. Overexpression causes a delay in G2 and is lethal in cdc13ᵗˢ mutants.	Forsburg, S.L. & Nurse, P. *Nature* **351**, 245–248 (1991).	YSPUC1
321	sct1	Encodes a relative of Cdc10. Dominant mutations in sct1 suppress cdc10ᵗˢ or Δcdc10 mutants. The double mutants have multiple cytoskeletal abnormalities. See cdc10.	Marks, J., Fankhauser, C., Reymond, A. & Simanis, V. *J Cell Sci* **101**, 801–808 (1992).	YSPSCT

Kinases and phosphatases

Entry	Gene	Description	Reference	Genbank Name
322	dis1	Encodes a protein that interacts with Ppa2 and has homology to a regulatory subunit of protein phosphatase 2A. cs mutants arrest in mitosis with sister chromatid pairs distributed throughout the spindle.	M. Yanagida, personal communication	

Entry	Gene	Description	Reference	Genbank Name
323	*dis2*	*dis2* and *dis21* both encode members of the protein phosphatase 1 family. A semidominant *dis2*^{cs} mutant arrests in mitosis with sister chromatid pairs distributed throughout the spindle. Double deletion of *dis2* and *sds21* is lethal and causes metaphase arrest.	Kinoshita, N., Ohkura, H. & Yanagida, M. *Cell* **63**, 405–415 (1990).	M27068
324	*dis3*	Encodes a weak homolog of budding yeast Ssd1. *cs* mutants arrest in mitosis with sister chromatid pairs distributed throughout the spindle. Interacts genetically with *dis1, dis2,* and *ppe1*.	Kinoshita, N., Ohkura, H. & Yanagida, M. *Cell* **63**, 405–415 (1990).	YSPDIS3P
325	*sds21*	See *dis2*.		M27069
326	*sds22*	Cloned but not homologous to known genes. A *ts* mutant arrests in metaphase. Sds22 binds to and alters the substrate specificity of phosphatase 1. Interacts genetically with *dis2*.	Stone, E.M. & Yanagida, M. *Curr. Biol.* **3**, 13–26 (1993).	YSPSDS22
327	*ppa1*	*ppa1* and *ppa2* both encode protein phosphatase 2A. Deletion of both genes is lethal, whereas Δ*ppa2* accelerates entry into mitosis.	Kinoshita, N., Ohkura, H. & Yanagida, M. *Cell* **63**, 405–415 (1990).	YSPPPA1
328	*ppa2*	See *ppa1*. Interacts genetically with *wee1, cdc25,* and *dis1*.		YSPPPA2
329	*ppe1*	Encodes a protein phosphatase that is closely related to budding yeast Sit4. Interacts genetically with *dis3, pck1, pim1, ppa1,* and *ppa2*. Δ*ppe1* cells are *cs* and have altered cell morphology at 26°C.	Shimanuki, M., *et al. Mol Cell Biol* **4**, 303–313 (1993).	D13712 (accession number)
330	*ppb1*	Encodes a protein with homology to protein phosphatase 2B. Δ*ppb1* cells are viable but have defects in cytokinesis.	M. Yanagida, personal communication	
331	*pck1*	*pck1* and *pck2* encode protein kinases related to protein kinase C. Neither gene is essential, but deletion of both is lethal. Δ*pck2* strains show abnormal cell morphology, and overexpression of *pck2* is lethal.	Toda, T., Shimanuki, M. & Yanagida, M. *EMBO J*, **in press** (1993).	
332	*pck2*	See *pck1*.		
333	*dsk1*	Encodes a serine, threonine, and tyrosine kinase. Overexpression suppresses *dis1*^{cs}. Activated by phosphorylation and enters the nucleus during mitosis.	Takeuchi, M. & Yanagida, M. *Mol Cell Biol* **4**, 247–260 (1993).	D13447 (accession number)

Entry	Gene	Description	Reference	Genbank Name
334	*pyp1*	*pyp1* and *pyp2* encode related tyrosine phosphatases. Neither gene is essential, but deletion of both is lethal. Genetic interactions suggest that Pyp1 and Pyp2 stimulate the activity of Wee1.	Ottilie, S., *et al. Mol Cell Biol* **12,** 5571–5580 (1992).	YSPPYP
335	*pyp2*	See *pyp1*.		
336	*pyp3*	Encodes a tyrosine phosphatase. Overexpression can suppress *cdc25*[ts], suggesting that Pyp3 can, like Cdc25, dephosphorylate Cdc2.	Millar, J.B.A., Lenares, G. & Russell, P. *EMBO J* **11,** 4933–4941 (1992).	

Replication functions

Entry	Gene	Description	Reference	Genbank Name
337	*pol1*	Encodes the catalytic subunit of DNA polymerase α. Essential.	Bouvier, D., *et al. Exp Cell Res* **198,** 183–190 (1992).	YSPPOL1G
338	*pol3*	Encodes the catalytic subunit of DNA polymerase δ. Essential.	Pignede, G., Bouvier, D., de, R.A. & Baldacci, G. *J Mol Biol* **222,** 209–218 (1991).	YSPPOL3
339	*pcn1*	Encodes the fission yeast homolog of PCNA (proliferating cell nuclear antigen), an accessory subunit of DNA polymerase δ. Essential.	Waseem, N.H., Labib, K., Nurse, P. & Lane, D.P. *EMBO J* **11,** 5111–5120 (1992).	SPPCNG
340	*nda1*	Encodes a homolog of budding yeast Mcm2.	M. Yanagida, personal communication	
341	*nda4*	Encodes a homolog of budding yeast Cdc46.	M. Yanagida, personal communication	

Mitotic components

Entry	Gene	Description	Reference	Genbank Name
342	*nda2*	Encodes an α tubulin (α1). *cs* mutants arrest in mitosis with disrupted spindles.	Adachi, Y., Toda, T., Niwa, O. & Yanagida, M. *Mol Cell Biol* **6,** 2168–2178 (1986).	YSPTUBA1
343	*nda3*	Encodes β tubulin. *cs* mutants arrest in mitosis with disrupted spindles.	Kanbe, T., Hiraoka, Y., Tanaka, K. & Yanagida, M. *J Cell Sci* **96,** 275–282 (1990).	YSPTUBB
344	*gtb1*	Encodes gamma tubulin, which is found at the spindle pole bodies. Essential.	Stearns, T., Evans, L. & Kirschner, M. *Cell* **65,** 825–836 (1991).	YSPGAMT
345	*abt2*	Encodes an α tubulin (α2). Not essential.	Adachi, Y., Toda, T., Niwa, O. & Yanagida, M. *Mol Cell Biol* **6,** 2168–2178 (1986).	YSPTUBA2

Entry	Gene	Description	Reference	Genbank Name
346	*cut1*	Encodes a protein with homology to budding yeast Esp1. *ts* mutants show abnormal coordination of septation formation and mitosis leading to guillotined nuclei, anucleate cells, and multinucleate cells. Interacts genetically with *cut2*, *cut4*, and *cut8*.	Uzawa, S., *et al. Cell* **62**, 913–925 (1990).	YSPCUT1
347	*cut2*	Cloned but not homologous to known genes. Mutant phenotype like *cut1⁻*.	Uzawa, S., *et al. Cell* **62**, 913–925 (1990).	YSPCUT2
348	*cut4*	Not cloned. Mutant phenotype like *nuc2^{ts}*.		
349	*cut7*	Encodes a member of the kinesin superfamily, and is closely related to *Aspergillus* BimC. Cut7 is found at or near spindle pole bodies. *ts* mutants form two unconnected half spindles.	Hagan, I. & Yanagida, M. *Nature* **356**, 74–76 (1992).	
350	*cut8*	Not cloned. Mutant phenotype like *nuc2^{ts}*.		
351	*cut9*	Encodes a homolog of budding yeast Cdc16. Mutant phenotype like *nuc2^{ts}*.	Goebl, M. & Yanagida, M. *Trends Biochem Sci* **16**, 173–177 (1991).	
352	*nuc1*	Encodes largest subunit of RNA polymerase I. *ts* mutants arrest in mitosis with an abnormal nucleolus.	Hirano, T., Konoha, G., Toda, T. & Yanagida, M. *J Cell Biol* **108**, 243–253 (1989).	YSPNUC1
353	*nuc2*	Encodes a protein that contains the loosely conserved repeat found in budding yeast Cdc16 and Cdc23. Nuc2 is tightly bound to the nuclear scaffold. *ts* mutants arrest in metaphase.	Hirano, T., Kinoshita, N., Morikawa, K. & Yanagida, M. *Cell* **60**, 319–328 (1990).	YSPNUC2
354	*sad1*	Encodes a protein located at the spindle pole body. Essential.	M. Yanagida, personal communication	
355	*top1*	Encodes topoisomerase I. Not essential. Cells require topoisomerase I or topoisomerase II activity during DNA replication.	Uemura, T., *et al. Nucleic Acids Res* **15**, 9727–9739 (1987).	YSPTOP1
356	*top2*	Encodes topoisomerase II. *ts* mutants are defective in chromosome condensation and sister chromatid separation in anaphase.	Shiozaki, K. & Yanagida, M. *J. Cell Biol.* **119**, 1023–1036 (1992).	YSPTOP2
357	*crm1*	Cloned but not homologous to known genes. *ts* mutants have disrupted chromosome structure. Crm1 is located in the nucleus. Interacts genetically with *pap1*, which encodes a transcription factor.	Toda, T., *et al. Mol Cell Biol* **12**, 5474–5484 (1992).	YSPCRM1

Entry	Gene	Description	Reference	Genbank Name

Feedback control

358	*rad1*	Cloned but not homologous to known genes. *rad1⁻*, *rad3⁻*, *rad9⁻*, and *rad17⁻* mutations inactivate the feedback control that keeps cells with damaged or unreplicated DNA from entering mitosis.	Rowley, R., Subramani, S. & Young, P.G. *Embo J* **11**, 1335–1342 (1992).	YSPRAD1
359	*rad3*	Cloned but not homologous to known genes. See *rad1*.	Seaton, B.L., Yucel, J., Sunnerhagen, P. & Subramani, S. *Gene* **119**, 83–9 (1992).	SPRAD3
360	*rad9*	Cloned but not homologous to known genes. See *rad1*.	Lieberman, H.B., Hopkins, K.M., Laverty, M. & Chu, H.M. *Mol Gen Genet* **232**, 367–376 (1992).	YSPRAD9
361	*rad17*	Not cloned but not homologous to known genes. See *rad1*.	al Khodairy, F. & Carr, A.M. *Embo J* **11**, 1343–1350 (1992).	
362	*rad21*	Not cloned. Mutations inactivate the feedback control that keeps cells with damaged DNA from entering mitosis.	al Khodairy, F. & Carr, A.M. *Embo J* **11**, 1343–50 (1992).	
363	*hus1*	Cloned but not homologous to known genes. *hus1⁻ and hus2⁻* mutants all inactivate the feedback control that keeps cells with damaged or unreplicated DNA from entering mitosis.	Enoch, T., Carr, A.M. & Nurse, P. *Genes Dev* **6**, 2035–2046 (1992).	
364	*hus2*	Not cloned. See *hus1*.		
365	*hus3*	Not cloned. Recessive *hus3⁻* and *hus4⁻* and dominant *hus5ᴰ* mutants increase sensitivity to unreplicated and damaged DNA by affecting feedback controls. *hus3⁻* is suppressed by growth on minimal medium.	Enoch, T., Carr, A.M. & Nurse, P. *Genes Dev* **6**, 2035–2046 (1992).	
366	*hus4*	Not cloned. See *hus3*.		
367	*hus5*	Not cloned.		
368	*pim1*	Also called *dcd1*. Encodes the fission yeast homolog of budding yeast Srm1 and human RCC1. *ts* mutants have been isolated that prevent chromosome decondensation as cells exit mitosis or allow cells with unreplicated DNA to enter mitosis.	Matsumoto, T. & Beach, D. *Cell* **66**, 347–360 (1991).	YSPRCCIB
369	*spi1*	Also called *fyt1*. Encodes the fission yeast homolog of mammalian TC4, the G protein whose function is regulated by RCC1. Overexpression suppresses mutations in *pim1*.	Matsumoto, T. & Beach, D. *Cell* **66**, 347–360 (1991).	YSPRCCIA

Entry	Gene	Description	Reference	Genbank Name
370	*cut5*	Also called *rad4*. Mutants are defective in DNA replication and are able to enter mitosis with unreplicated DNA.	M. Yanagida, personal communication	YSPRAD4

Mating and meiosis

Entry	Gene	Description	Reference	Genbank Name
371	*mat1-Pc*	The *mat1-P* mating-type locus found in h^+ strains contains two genes, *mat1-Pc* and *mat1-Pi*. *mat1-Pc* is required for both mating and sporulation. Full expression requires nitrogen starvation.	Nielsen, O., Davey, J. & Egel, R. *Embo J* 11, 1391–1395 (1992).	YSPMAT1P
372	*mat1-Pi*	Also called *mat1-Pm*. Encodes a protein with homology to transcription factors. Required only for sporulation. Expression requires nitrogen starvation and M factor.	Nielsen, O. & Egel, R. *Embo J* 9, 1401–1406 (1990).	YSPMAT1P
373	*mat1-Mc*	The *mat1-M* mating type locus found in h^- strains contains two genes, *mat1-Mc* and *mat1-Mi*. *mat1-Mc* is required for both mating and sporulation. Full expression requires nitrogen starvation.	Kelly, M., *et al*. *Embo J* 7, 1537–1547 (1988).	YSPMAT1M
374	*mat1-Mi*	Also called *mat1-Mm*. Encodes a protein with similarities to non-histone chromatin proteins. Required only for sporulation. Expression requires nitrogen starvation.	Kelly, M., *et al*. *Embo J* 7, 1537–1547 (1988).	YSPMAT1M
375	*mam2*	*mam2* encodes the receptor for P mating factor and *map3* encodes the receptor for M mating factor. Receptor expression is induced by nitrogen starvation. Either the P-factor or M-factor signaling pathway is sufficient to induce meiosis and sporulation.	Kitamura, K. & Shimoda, C. *Embo J* 10, 3743–3751 (1991).	YSPMAMPR
376	*map3*	See *mam2*.	Tanaka, K., Davey, J., Imai, Y. & Yamamoto, M. *Mol Cell Biol* 13, 80–88 (1993).	YSPMFACR
377	*gpa1*	Encodes the α subunit of a trimeric G protein. Essential for mating and sporulation, but not for the mitotic cell cycle. Thought to be coupled to mating factor receptors. Activated alleles can induce the mating response in nitrogen-starved cells that lack mating partners.	Obara, T., Nakafuku, M., Yamamoto, M. & Kaziro, Y. *Proc Natl Acad Sci U S A* 88, 5877–81 (1991).	YSPGPA1

Entry	Gene	Description	Reference	Genbank Name
378	*ras1*	Also called *ste5*. Encodes Ras. Δ*ras1* cells are sterile and sporulate poorly, whereas activated alleles increase mating factor sensitivity. Even in the presence of activated Ras, mating still requires nitrogen starvation and mating factors.	Nielsen, O., Davey, J. & Egel, R. *Embo J* **11**, 1391–1395 (1992).	YSPRASX
379	*gap1*	Also called *sar1*. Encodes a GTPase activating protein that stimulates GTP hydrolysis by Ras.	Imai, Y., Miyake, S., Hughes, D.A. & Yamamoto, M. *Mol Cell Biol* **11**, 3088–3094 (1991).	YSPGAP1
380	*ste6*	Encodes the homolog of the budding yeast Cdc25. Essential for mating but not for the mitotic cell cycle.	Hughes, D.A., Fukui, Y. & Yamamoto, M. *Nature* **344**, 355–357 (1990).	YSPST6
381	*ral2*	Sequenced but shows no homology to known genes. Δ*ral2* cells proliferate normally but fail to mate unless they contain a dominant activated *ras1*[D] mutant.	Fukui, Y., Miyake, S., Satoh, M. & Yamamoto, M. *Mol Cell Biol* **9**, 5617–5622 (1989).	YSPRAL2
382	*byr1*	Also called *ste1*. Encodes a homolog of budding yeast Ste7 and vertebrate MAP kinase kinase. Byr1 is activated by Byr2 and probably phosphorylates and activates Spk1. Overexpression allows Δ*ras1* cells to sporulate.	Nadin, D.S. & Nasim, A. *Mol Cell Biol* **10**, 549–560 (1990).	YSPBYR1
383	*byr2*	Also called *ste8*. Encodes a homolog of budding yeast Ste11. Downstream of Ras1 in the mating factor signaling pathway. Byr2 may phosphorylate and activate Spk1. Overexpression allows Δ*ras1* cells to sporulate.	Wang, Y., et al. *Mol Cell Biol* **11**, 3554–3563 (1991).	YSPBYR2
384	*byr3*	Encodes a transcription factor essential for efficient mating but dispensable for the mitotic cell cycle. Overexpression allows Δ*ras1* cells to sporulate.	Wang, Y., *et al. Mol Cell Biol* **11**, 3554–3563 (1991).	S112402
385	*spk1*	Encodes a member of the MAP kinase family similar to Fus3. Required for mating but not for the mitotic cell cycle.	Toda, T., Shimanuki, M. & Yanagida, M. *Genes Dev* **5**, 60–73 (1991).	YSPSPK1
386	*gpa2*	Encodes the α subunit of a trimeric G protein. Δ*gpa2* cells have reduced cyclic AMP levels and can mate and sporulate on rich medium. Probably regulates adenyl cyclase in response to nutritional signals.	Isshiki, T., Mochizuki, N., Maeda, T. & Yamamoto, M. *Genes & Dev.* **6**, 2455–2462 (1992).	D13366

Entry	Gene	Description	Reference	Genbank Name
387	*cap1*	Encodes the homolog of budding yeast Cap. Deletion increases mating and sporulation, whereas overexpression leads to sterility.	Kawamukai, M., *et al. Mol Biol Cell* **3**, 167–180 (1992).	
388	*cyr1*	Encodes adenyl cyclase. Not essential. Δ*cyr1* cells can mate on rich medium.	Maeda, T., Mochizuki, N. & Yamamoto, M. *Proc Natl Acad Sci U S A* **87**, 7814–7818 (1990).	YSPADC
389	*pde1*	Also called *cgs2*. Encodes cyclic AMP phosphodiesterase. Deletions make cells partially sterile and block meiosis.	Mochizuki, N. & Yamamoto, M. *Mol Gen Genet* **233**, 17–24 (1992).	
390	*pka1*	Encodes the catalytic subunit of cyclic AMP-dependent kinase. Overexpression causes sterility.	T. Maeda and M. Yamamoto, personal communication	
391	*cgs1*	Encodes the regulatory subunit of cyclic AMP-dependent kinase. Mutants are sterile and cannot enter G0.	DeVoti, J., Seydoux, G., Beach, D. & McLeod, M. *Embo J* **10**, 3759–3768 (1991).	S64907
392	*pat1*	Also called *ran1*. Encodes a protein kinase whose activity is required to keep cells in the mitotic cell cycle. Mei3 binds to and inactivates Pat1 in meiotic cells.	Nielsen, O. & Egel, R. *Embo J* **9**, 1401–1406 (1990).	YSPRAN1G
393	*ste11*	Encodes a transcription factor required for transcription of many genes involved in mating and sporulation. Expression requires nitrogen starvation and a decrease in the cyclic AMP level.	Sugimoto, A., *et al. Genes Dev* **5**, 1990–1999 (1991).	YSPSTE11
394	*mei2*	Encodes a gene that is required for meiosis. Inactivation of Pat1 leads to activation of Mei2 and induction of meiosis. Expression requires Ste11 and can be blocked by high levels of cyclic AMP.	Sugimoto, A., *et al. Genes Dev* **5**, 1990–1999 (1991).	YSPMEI2
395	*mei3*	Encodes a protein kinase inhibitor. Expression requires Ste11 and the mating factor signaling pathway. See *pat1*.	McLeod, M. & Beach, D. *Nature* **332**, 509–514 (1988).	YSPMEI3G
396	*pac1*	Encodes a double-stranded ribonuclease. Overexpression prevents mating and meiosis in wild-type and *ran1*[ts] cells.	Iino, Y., Sugimoto, A. & Yamamoto, M. *Embo J* **10**, 221–226 (1991).	YSPPAC1G
397	*ste4*	Encodes a transcription factor required for mating and meiosis, whose expression requires the activity of Ste11.	Okazaki, N., Okazaki, K., Tanaka, K. & Okayama, H. *Nucleic Acids Res* **19**, 7043–7047 (1991).	SPSTE4

Entry	Gene	Description[a]	Reference	Genbank Name
398	*mfm1*	*mfm1* and *mfm2* both encode the precursor for M factor, which like budding yeast **a** factor is prenylated.	Davey, J. *Embo J* **11**, 951–960 (1992).	S87829
399	*mfm2*	See *mfm1*.		S87831
400	*sxa1*	Encodes a protease that probably degrades M factor. h^+ $\Delta sxa1$ cells are hypersensitive to mating factor and mate poorly.	Imai, Y. & Yamamoto, M. *Mol Cell Biol* **12**, 1827–1834 (1992).	YSPSXA1
401	*sxa2*	Encodes a protease that probably degrades P factor. h^- $\Delta sxa2$ cells are hypersensitive to mating factor and mate poorly.	Imai, Y. & Yamamoto, M. *Mol Cell Biol* **12**, 1827–1834 (1992).	YSPSXA2

ASPERGILLUS

Entry	Gene	Description	Reference	Genbank Name
402	*nimA*	Encodes a protein kinase whose activity is required for entry into mitosis. $bimE^{ts}$ cells can enter mitosis in the absence of detectable NimA associated kinase, but spindle formation is abnormal.	Osmani, A.H., McGuire, S.L. & Osmani, S.A. *Cell* **67**, 283–291 (1991).	EMENIMA
403	*nimE*	Encodes a B-type cyclin. *ts* mutants block entry into mitosis. Mild overexpression can suppress $nimT^{ts}$ mutations.	O'Connell, M.J., Osmani, A.H., Morris, N.R. & Osmani, S.A. *Embo J* **11**, 2139–2149 (1992).	EMENIMEMR
404	*nimT*	Encodes a homolog of fission yeast Cdc25. *ts* mutants block entry into mitosis.	O'Connell, M.J., Osmani, A.H., Morris, N.R. & Osmani, S.A. *Embo J* **11**, 2139–2149 (1992).	EMENIMTMR
405	*bimA*	Encodes a protein with homology to fission yeast Nuc2 and budding yeast Cdc23. BimA localizes to the spindle pole body. *ts* mutants arrest in mitosis.	O'Donnell, K.L., Osmani, A.H., Osmani, S.A. & Morris, N.R. *J Cell Sci* **199**, 711–719 (1991).	EMEBIMA
406	*bimB*	Encodes a protein with homology to budding yeast Esp1 and fission yeast Cut1. A *ts* mutant forms a transient spindle but does not perform nuclear division and undergoes multiple rounds of DNA replication.	May, G.S., McGoldrick, C.A., Holt, C.L. & Denison, S.H. *J Biol Chem* **267**, 15737–15743 (1992).	EMEBIMB
407	*bimC*	Encodes a heavy chain of the kinesin family with strong homology to fission yeast Cut7. A *ts* mutation is defective in spindle pole body separation and nuclear division. This defect is suppressed by $\Delta klpA$.	O' Connell, M.J., Meluh, P.B., Rose, M.D. & Morris, N.R. *J. Cell Biol.* **120**, 153–162 (1993).	EMEBIMC4A

Entry	Gene	Description	Reference	Genbank Name
408	*bimD*	Encodes a potential DNA-binding protein. *ts* mutants have defective spindles at 42°C and are hypersensitive to DNA damage at 25°C. Overexpression leads to cell cycle arrest in G1 or S phase.	G. May, personal communication	
409	*bimE*	Encodes a predicted transmembrane protein. *ts* mutants enter mitosis from any point in the cell cycle when raised to 42°C.	Osmani, A.H., O'Donnell, K., Pu, R.T. & Osmani, S.A. *Embo J* **10**, 2669–2679 (1991).	EMEBIME
410	*bimG*	Encodes protein phosphatase 1. A *ts* mutant shows incomplete anaphase and a high fraction of mitotic cells.	Doonan, J.H. & Morris, N.R. *Cell* **57**, 987–996 (1989).	M27067
411	*klpA*	Encodes a homolog of the budding yeast Kar3 protein. Overexpression blocks spindle function. Not essential.	O' Connell, M.J., Meluh, P.B., Rose, M.D. & Morris, N.R. *J Cell Biol* **120**, 153–162 (1993).	EMEKLPAMR
412	*benA*	Encodes a β tubulin. Essential.	Jung, M.K., Wilder, I.B. & Oakley, B.R. *Cell Motil Cytoskeleton* **22**, 170–174 (1992).	EMETUBBA
413	*tubA*	Encodes an α tubulin that is essential for nuclear migration and division.	Doshi, P., *et al. Mol Gen Genet* **225**, 129–141 (1991).	
414	*tubB*	Encodes an α tubulin. Not required for the vegetative life cycle, but Δ*tubB* cells block early in meiosis.	Kirk, K.E. & Morris, N.R. *Genes Dev* **5**, 2014–2023 (1991).	
415	*tubC*	Encodes a β tubulin that is expressed during spore formation.	May, G.S., Waring, R.B. & Morris, N.R. *Cell Motil Cytoskeleton* **16**, 214–220 (1990).	EMETUBBB
416	*mipA*	Encodes γ tubulin. Essential for spindle formation and nuclear division. Spores that germinate without γ tubulin do not arrest in mitosis. Interacts genetically with *benA*.	Oakley, B.R., Oakley, C.E., Yoon, Y. & Jung, M.K. *Cell* **61**, 1289–1301 (1990).	EMEMIPAG
417	*nudC*	Cloned but not homologous to known genes. *ts* mutants block nuclear migration.	Osmani, A.H., Osmani, S.A. & Morris, N.R. *J Cell Biol* **111**, 543–551 (1990).	EMENMP
418	*camA*	Encodes calmodulin. Essential for entry into mitosis.	Rasmussen, C.D., *et al. J Biol Chem* **265**, 13767–13775 (1990).	

INDEX TO GENES THAT AFFECT THE CELL CYCLE

This list is an index to the table of gene descriptions on pages 201–239. The number in the right column is the number of the entry in the table.

Gene Name	Entry	Gene Name	Entry	Gene Name	Entry
cdr1	312	CYR1	30	HOP1	272
cdr2	313	cyr1	388	HRR25	118
cgs1	391			HTA1	147
cgs2	389	DAC2	235	HTA2	148
CHL1	177	DAF1	57	HTB1	149
CHL15	125	DAF2	238	HTB2	150
cig1	319	DBF2	119	hus1	363
CIK1	173	DBF20	120	hus2	364
CIN1	161	DBF4	146	hus3	365
CIN2	162	dcd1	368	hus4	366
CIN4	163	DIN1	132	hus5	367
CIN8	166	DIS1	277		
CIN9	167	dis1	322	IME1	281
CKA1	114	DIS2	95	IME2	282
CKA2	115	dis2	323	IME4	283
CKS1	67	dis3	324	IRA1	83
CLB1	59	DMC1	265	IRA2	84
CLB2	60	DNA33	50		
CLB3	61	DPB2	126	KAR1	164
CLB4	62	DPB3	127	KAR3	165
CLB5	63	DPR1	86	KEM1	266
CLB6	64	dsk1	333	KEX1	244
CLN1	55			KEX2	245
CLN2	56	ESP1	171	KIP1	167
CLN3	57			KIP2	168
CLS4	22	FAR1	237	klpA	411
CMD1	108	FUS1	233	KSL2	166
CMK1	109	FUS2	234	KSS1	236
CMK2	110	FUS3	235		
CMP1	105	fyt1	369	MAD1	209
CMP2	106			MAD2	210
CNA1	105	GAC1	96	MAD3	211
CNA2	106	gap1	379	mam2	375
CNB1	107	GLC7	95	map3	376
crm1	357	GPA1	231	mat1-Mc	373
CTF4	125	gpa1	377	mat1-Mi	374
CTF13	175	gpa2	386	mat1-Mm	374
CTF14	176	gtb1	344	mat1-Pc	371
cut1	346			mat1-Pi	372
cut2	347	HCS26	58	mat1-Pm	372
cut4	348	HHF1	153	MATα1	212
cut5	370	HHF2	154	MATa1	214
cut7	349	HHT1	151	MATα2	213
cut8	350	HHT2	152	MCK1	112
cut9	351	HO	216	MCM1	215

Gene Name	Entry	Gene Name	Entry	Gene Name	Entry
MCM2	133	*nuc1*	352	*PRP20*	241
MCM3	134	*nuc2*	353	*PRT1*	49
MCM5	39	*nudC*	417	*PTP1*	92
mcs1	317			*PTP2*	93
mcs4	318			*PTP2*	94
MEC1	203	*ORC1*	139	*puc1*	320
MEC2	204	*ORC2*	140	*PYK1*	18
MEC3	205	*ORC3*	141	*pyp1*	334
mei2	394	*ORC4*	142	*pyp2*	335
mei3	395	*ORC5*	143	*pyp3*	336
MEI4	270	*ORC6*	144		
MEK1	274			*rad1*	358
MER1	268	*pac1*	396	*rad3*	359
MER2	269	*pat1*	392	*rad4*	370
MFA1	247	*PBS2*	113	*RAD9*	200
MFα1	242	*pck1*	331	*rad9*	360
MFA2	248	*pck2*	332	*RAD17*	201
MFα2	243	*pcn1*	339	*rad17*	361
mfm1	398	*PDE1*	75	*rad21*	362
mfm2	399	*pde1*	389	*RAD24*	202
MIH1	65	*PDE2*	76	*RAD50*	260
mik1	311	*PFY1*	79	*RAD51*	261
mipA	416	*pim1*	368	*RAD52*	262
MPB1	70	*pka1*	390	*RAD54*	263
MPS1	169	*PKC1*	111	*RAD57*	264
MPS2	170	*PKX2*	4	*ra12*	381
MRE4	274	*PMS1*	279	*RAM1*	85
MSB1	185	*POB1*	125	*RAM2*	86
MSB2	186	*POL1*	16	*ran7*	392
MSI1	88	*pol1*	337	*RAS1*	80
MYO1	195	*POL2*	122	*ras1*	378
MYO2	196	*POL3*	1	*RAS2*	81
MYO4	197	*pol3*	338	*REC102*	271
		POL30	124	*RED1*	275
nda1	340	*ppa1*	327	*RED2*	276
nda2	342	*ppa2*	328	*RFA1*	136
nda3	343	*ppb1*	330	*RFA2*	137
nda4	341	*ppe1*	329	*RFA3*	138
NDC1	172	*PPH21*	97	*RME1*	280
NDC10	176	*PPH22*	98	*RNR1*	130
nim1	312	*PPH3*	99	*RNR2*	131
nimA	402	*PPZ1*	103	*RNR3*	132
nimE	403	*PPZ2*	104	*RPA1*	136
nimT	404	*PRI1*	128	*RPI1*	89
NMT1	53	*PRI2*	129	*RSR1*	178

Gene Name	Entry	Gene Name	Entry	Gene Name	Entry
SAC6	194	*SST1*	246	*top1*	355
sad1	354	*SST2*	239	*TOP2*	156
sar1	379	*ste1*	382	*top2*	356
SCD25	82	*STE2*	218	*TOP3*	157
SCG1	231	*STE3*	219	*TPD3*	100
SCG2	86	*STE4*	220	*TPK1*	71
SCH9	90	*ste4*	397	*TPK2*	72
sct1	321	*STE5*	221	*TPK3*	73
sds21	325	*ste5*	378	*TPM1*	199
sds22	326	*STE6*	222	*TUB1*	158
SEP1	266	*ste6*	380	*TUB2*	159
SGV1	240	*STE7*	223	*TUB3*	160
SID1	278	*ste8*	383	*tubA*	413
SIT4	101	*STE11*	224	*tubB*	414
SLK1	188	*ste11*	393	*tubC*	415
SME1	282	*STE12*	225		
SMY2	198	*STE13*	226	*VPS1*	258
SNC1	87	*STE14*	227		
SPA2	187	*STE16*	86	*wee1*	310
spi1	369	*STE18*	229	*WHI1*	57
spk1	385	*STE20*	230	*wis1*	315
SPO7	253	*stf1*	314		
SPO11	254	*suc1*	316	*XRN1*	266
SPO12	255	*SWE1*	66	*XRS2*	267
SPO13	256	*SWI4*	68		
SPO14	257	*SWI5*	217	*YAK1*	91
SPO15	258	*SWI6*	69	*YCK1*	116
SPO16	259	*sxa1*	400	*YCK2*	117
SPT16	52	*sxa2*	401	*YPK1*	112
SRM1	241				
SSD1	102	*TMP1*	20	*ZIP1*	273
SSL1	249	*TOP1*	155		

INDEX